Royal Air Observation Post
Auster Era
1936-1956
Evolution & Campaigns

Major Aamir Mushtaq Cheema (Retired)
Master of Letters, War Studies (Glasgow)

majoraamircheema@yahoo.co.uk
Tel: +92 (0)302 547 8100

Designer: Syed Afraz Ali Kazmi

Royal Air Observation Post 1936-1956

Published in Islamabad, Pakistan in 2012

Now Reprinted March 5, 2018
by Ishi Press in New York and Tokyo

Ishi Press International
1664 Davidson Ave, Suite 1B
Bronx NY 10453-7877
USA
917-659-3397
917-507-7226
samhsloan@yahoo.com
samhsloan@gmail.com

ISBN 4-87187-180-0
978-4-87187-180-8
Printed in the United States of America

For
Salik & Azadeh

Royal Air Observation Post

Auster Era

1936-1956

Evolution & Campaigns

First Published in Pakistan 2012
by Verve Communication
ISBN 9789699246012

Author. Major Aamir Mushtaq Cheema (Retired)
majoraamircheema@yahoo.co.uk
Tel: +92 (0)302 547 8100

Verve Communication/Printers/Publishers
Suite # 11, Block # 15,
Street # 40, I&T Center, G-10/4,
Islamabad, Pakistan
vervecommunication@yahoo.com
Tel: +92 (0)51 221 5302

Designer: Syed Afraz Ali Kazmi

Foreword

Brigadier Malik Mukhtar Karim {retired}

Brigadier MM Karim was born in 1925 graduate of Prince of Wale's Royal Military College Dehra Dun in 1943, commissioned in Royal Indian Air Force as pilot in 1944,he later served as Midshipman in Royal Indian Navy in 1946 and finally passed out from Indian Military Academy Dehra Dhun in 1947 with a Sword of Honour in an artillery regiment.In Pakistan he served as Aide De Camp to Commander–in-Chief Pakistan Army General Sir Douglas Gracey,earned his flying brevet as an Air Observation Pilot in 1950,raised the very first Air Observation Squadron in 1956, commanded the Army Aviation School in 1965 and after his retirement became the pioneer hot air balloonist of Pakistan, he is still living an active life in Islamabad and I pray for him to complete his life century.

At the long finals of my life when I look back I have nothing but gratitude for an exciting life which I lived thanks to the uniforms. Military life is a wonderful experience which every young man must have a flavor of it, nothing is more sacred to me than serving the country. Flying is a thrill and nothing matches it in this world, to be able to see the world from top is an exhilarating phenomena. I was lucky to wear the uniform of three services and as such I have found that the military culture varies in them the way of working and the mind set are different in them; the concept of espirit de corps also differs. Coming to the Air Observation Post, basing upon my personal experience of the abinitio days in Pakistan I think we{Air Observation Pilots} benefitted from the first two commander in chiefs who were British and later followed by General Ayub Khan the very first Pakistani commander in chief, on the air force side the role of Air Chief Marshall Asghar Khan is immense in the formation of army flying. I agree that there exist a friction among all the three services and this phenomenon is world over something very peculiar to military; however in case of Pakistan the induction of American military equipment in 1954 onwards drastically change it. This created an infusion of three distinct military cultures the British, American and native Pakistani which was still trying to find its roots separately from the India. The sense of self sacrificing in almost all the newly born nations is more powerful than the hereditary military traditions.

In Pakistan the Air Observation Post was much more advance than the neighbouring India, we were the first one to induct non gunner pilots in it and one of the pioneer countries to have an independent army aviation in 1977, one of the reason for this is the personal and untiring efforts of Lieutenant General Azmat Baksh Awan {Late}. The point which I want to drive home is that power of personality plays an important role in defining historical events, within the military system which in my opinion is no different from the model of bureaucracy the things can either move very fast if the individuals are in a rhythm or vice versa it can take as long as one can imagine. The raising of Air Observation Post Squadrons in United Kingdom is a classical example of both types of working.

In my opinion the key factor is the role of a pilot in the military system, how to treat him; he is a breed apart in himself and as such requires dealing which has to be different from the way the others are dealt with and in this aspect it is only air force which fully understands the psyche of an aviator. Army and Navy have different standards, traditions, customs and rules which in majority of cases are not very conducive for flyers.

I do not for see in near future that this issue can be resolve, the creation of army flying corps all over the world is the right step yet within them fundamentally the flyers are still governed by the ground rules. I can recall many an instances when the dress of a pilot was the subject of many long hours of discussions at various tiers of command & staff in United Kingdom, India and Pakistan military establishments at times the mere raising of collars by the aviators was frowned upon.

I am sure everyone will benefit from reading this book, it is rich in historical data and the abundant use of primary sources haves brought forward a new angle to the existing dimension on the subject. My best wishes are for all the Air observation Pilots all over the world; they in any case are now a rare breed.

Contents

Abbreviations and explanation of terms used

Ack Ack Anti Aircraft Artillery.

A.G.R.A Army Group Royal Artillery, a grouping of artillery regiments that include field, medium and heavy regiments, allocated to army .

A.O.P. Air Observation Post, a light unarmed aircraft flown exclusively by the Royal Artillery Officers.

Air OP Squadrons A Royal Air Force Unit consisting of Squadron Headquarters and three Flights 'A' 'B' & 'C' attached to the Royal Artillery of a corps.

A.L.G. Advance Landing Ground selected close to the forward defence line.

ALFSEA Allied Land Forces South East Asia.

BRA Brigadier Royal Artillery.

BMRA Brigade Major Royal Artillery in charge of Artillery at Divisional Headquarters.

Battery Sub section of an artillery regiment, 2-3 troops form a battery commanded by a major, it normally had 12 guns till 1940.

CFI Chief Flying Instructor. s

C.R.A Commander Royal Artillery, the appointment at divisional level of senior artillery officer, he was an advisor and not a commander and as such does not command directly the artillery regiments in the division

C.C.R.A. Commander Corps Royal Artillery, the senior artillery officer at corps level

Direct Fire. The artillery fire in which guns can see the target.

EFTS Elementary Flying training School.

Field, Medium and Heavy Artillery. The division of guns basing upon their range and calibre, Thus field guns had a range of 12000 yards, medium of 20000 yards and heavy more than these two.

F.O.O. Forward observer officer, an artillery officer who is attached

with the infantry battalion to carry out observed fire of his battery.

FOP Flying Observation Post it was later changed into Air Observation Post

G Staff. The staff in the divisional ,corps and at army level which looks after the actual operational requirement ,it is different from the 'Q' staff which looks after the administrative aspect.

GF Gunfire, Guns from artillery units fire specific number of rounds on to a target

HT Cables High Tension Cables

HQRA Headquarters Royal Artillery

Indirect Fire . Artillery fire in which guns rely upon the maps for firing.

IAF Indian Air Force

OP Observation Post.

OR All ranks other than Officers, Warrant Officers & Non Commissioned Officers.

OCU Officers Conversion Unit.

Ranging The procedure of adjusting the fall of artillery shot on to a specific spot before the actual fire.

RAF Royal Air Force The force in United Kingdom which controlled all flying it was formed in 1918 out of the army.

RFC Royal Flying Corps

RPAF Royal Pakistan Air Force ,the prefix royal was removed in 1956 .

SITREP Situation Report.

SEAC South East Asia Command.

Tac 21 Army Group The Tactical Headquarters of Field Marshal Montgomery.

VE Victory in Europe. 8th May 1945

VJ Victory over Japan,15th August 1945.

Introduction

Guns symbolize the battle by virtue of their firepower, thunder, roar and sheer size. In England the artillery was first employed in the battle of Crecy in 1346, there is a historical and academic debate about the origin of the gunpowder, to some it were the Arabs and to others they were the Chinese 'thunder of the earth' (an effect produced by filling a large bombshell with a gunpowder mixture) sounded faint reverberations amongst the philosophers of the western world as early as A.D. 300[1]. The young Arab general Mohammad Bin Qasim did employ Manjneeq a piece of sling with which he threw the heavy stones 'can you break the banner that flies at the head of that temple with your catapult, inquired Qasim from Juana Salmi his catapult operator'[2] on the Hindu Raja Dahir's Fort Deybul, Sind in 712 AD.As a matter of fact it was the David who first used the long range tactics when he hit Goliath with stones using his sling but the first show of firearms in Western Europe may have been by the Moors, at Saragossa, in A.D. 1118. In later years the Spaniards turned the new weapon against their Moorish enemies at the siege of Cordova (1280) and the capture of Gibraltar (1306).[3] Holy Roman Emperor Maximillian 1 in 1490 placed gunners as a special cadre in his army. In 1537 Italian mathematician Nicolo Fontana published first scientific treatise on artillery proving that maximum range is obtained when guns fire at an angle of 45 degrees and Spain in 1590 produced first artillery firing tables. King Charles V {1500-1558}and King Henry 2nd {1459-1519} standardised the artillery into following, Culvera{guns} long barrelled accurate & long range, Cannons{howitzers}short barrel,

1 Albert Manuq,*Artillery Through the ages,a short illustrated history of Cannon emphasizing types used in America*{US Government Press,Washington,1949},p-9.

2 Manan Ahmed,*Many histories of Muhammad bin Qasim narrating the conquest of Sind*,unpublished thesis for Doctor of Philosophy degree {University of Chicago,Illincis,2008},p-131.

3 Manuq,Artillery through ages,p-16.

heavy projectile & short distance, Pederero{mortars} short barrel, short distance & light projectile.

In 1670 the very first artillery company was raised in England however it was in France that Artillery reached its zenith first in the time of Charles 14th who separated artillery from the rest of army to give them their unique importance. Adolphus Gustavo's {1594-1632} in thirty years wars did brought a military revolution in terms of organisation and tactics including employing artillery and infantry together apart from standardizing the shell weight and artillery into various branches he had 12 pounder for field artillery and 24 pounder for siege works, not the least it was he who put gunners under army discipline one of his major achievement was in allowing the infantry and artillery to fight interdependently. It was again French who categorized artillery into field, siege, howitzer and mortar. Royal Regiment of Artillery after its raising with two permanent companies in 1716 virtually set the pace, trends and standards for British Army, they were the first one to have their own academy in 1746. They overcame their limitation of speed by raising Royal Horse Artillery. In January 1793 two troops of horse artillery were raised because as gunners had to walk while officers were mounted thus now all were mobile, each troop had six, six pounder guns and 45 drivers along with 187 horses 'a self contained mobile fighting unit of artillery had at last come into existence'. Till 1854 {Crimean War} artillery remained under the Royal Ordnance whose Master General Ordnance was also the ex-officio Colonel Commandant of Royal Artillery as well an artillery select committee was formed in 1805 to assist the master general. In 1807 during the battle of Friedland the French Artillery under Napolean pushed guns to within 60 yards of Prussian lines to support own infantry and cavalry. Each company or troop of RA was commanded by a captain who had another captain as his second-in-command at times instead of a captain, two or more lieutenants were posted,each battalion had two full colonels and a colonel commandant, it was the discretion of the colonel commandant to appoint the adjutant. the rank of regiment major was abolished in 1827 thus captains were promoted directly as lieutenant colonels. The artillery companies were at times designated as field batteries or brigades of guns. Battalion guns were grouped in pairs and three pairs were called a brigade,the term field artillery was introduced in 1879. Woolwich was the heart and soul of artillery. Each battalion made up of eight companies and one lieutenant colonel commanded two companies. The command of a battalion was always given to the senior most officer and if there were two officers of equal seniority then the senior most was given the command and junior posted out.Since 1821 every recruit in RA had been enlisted as 'gunner driver' the length of beard for a gunner was restricted to two fingers length and moustaches were made optional in July 1854.

Italians in 1846 devise the first rifled breech loading cannon with a range upto 4000 yards later Krupp{1812-1887} built rifled, steel, breech loading artillery.A Romanian Karl Gluk{1846-1910} laid the foundation of effective indirect artillery fire in 1882,with the help of compass and utilizing aiming points to direct artillery fire.

In 1889 RHA were lettered and field battery was numbered,in 1900 both types were amalgamated to form a 'brigade division' it was in 1924 that a single artillery corps was raised by putting RHA, field artillery and Royal Garrison Artillery together. By 1927 all medium artillery was tractor drawn, by 1937 nine field brigades were mechanized. The training memorandum of artillery in 1939 stressed 'The type of war we must now consider is not the highly mobile type but a more ponderous type of masses'.[4] In 1939 in place of brigades of four, six gun batteries; regiment of two batteries each of three troops of four guns

4 C.A.L.Grahaam, Brigadier General. *Story of Royal Regiment of Artillery*{Royal Artillery Institution,Woolwich,1962},p-70

were introduced, it was again change into three batteries each having two, four gun troops in 1940.[5] Troop is an old title of RHA for four guns; from its earliest days the Royal Horse Artillery enjoys precedence over all other regiments including the household cavalry.[6]

It was Fredrick the Great who introduced horse artillery in 1759. One of the first appearances of organized American field artillery on the battlefield was in the Northeast, where France's Louisburg fell to British and Colonial forces in 1745 yet there seems to be a shroud on the very nature of gunners as part of fighting arms, it was in the age of smooth bore 14th –late nineteen century in which artillery develop from novelty on battlefield to becoming battle winner mainly due to its ability to reduce siege warfare 'it emerged as an arm of equal status to the infantry and cavalry'[7] as late as the American Revolution there was a dispute about whether a general officer whose service had been in the Royal Artillery was entitled to command troops of all arms. There was no such question in England: the artillery general was a personage having 'always a part of the charge, and when the chief general is absent, he is to command all the army.'[8] English field artillery of the day had 'brigades' of four to six cannon, and each piece was supplied with 100 rounds of solid shot and 30 rounds of grape.

Chinese General Han Hsin 200 B.C., is reported to have flown the kite first for military purpose that is to assess the accurate distance between his forces and besieged fortress, through length of rope. Orville Wright took off in his 'Flyer' on 7 December 1903 at 10.35 for a flight of 120 feet in 12 seconds. The first powered, sustained and controlled flight in human history, although many others had been air borne in one way

5 Robert Lieutenant Colonel, ed, *Some Brief Reference Notes on the RHA*{March 1956}Fire Power Museum.
6 *Right of Line*,pp,11-57.
7 Unpublished thesis artillery warfare, Cranfield university,2008.p-34
8 Albert Manucy, Artillery Through the Ages,A Short Illustrated History of Cannon, Emphasizing Types Used in America, Superintendent of DocumentsU. S. Government Printing OfficeWashington , D. C.,1949, p-11.

or the other. It were the French who in 1783 first ascended in a balloon and later On 26 June 1794, in the Battle of Fleurus, Captain Coutelle of France ascended in balloon Entrepreneunt for directing the artillery fire, it is the first recorded event where third dimension was utilized by the land forces more specifically the gunners for their profession, soon Napoleon did made use of these balloons but later discarded them due to their heavy logistic support required, Napoleon was the first Gunner Emperor, President General Pervaz Musharraf of Pakistan (2000-2007) was also a gunner. Balloons were again utilized by the Unionist Forces in American Civil War; as a matter of fact both belligerents employed them for partial success. In Great Britain the first balloon company was raised as part of Rcyal Engineers in 1854, in 1898 a Polish scholar Ivan Bliokh published a study predicting that soon balloons will be used to drop explosive substance.'[9]

But what was the military purpose of the aircraft or airships. At the beginning of the last century there were two distinct schools of thoughts one favouring the airships which were more predominant before the Flyers flight, most notable among them was Count Zeppelin (1838-1917) of Germany and other were in favour of aircraft. Italians first used the aircraft in Libya when they bombed the Turks with hand thrown twenty kilogram odd bombs, military damage was negligible yet the morale factor which Napoleon rated as 3 to 1 in respect to material was justified. There seems little evidence that any of the belligerents on Europe had a precise idea of how to utilize this new weapon, which till then was not even considered as a weapon. In Germany the aircraft were all part of the military transport directorate, in France under the artillery and in Britain they were under the Royal Engineers. Their basic purpose at that time was mainly envisaged as the means of reconnaissance, they were to be the eyes of the field force; for the advantage of looking

9 Lee Kennett, *The First AirWar 1914-1918* {Free Press,NY,1991},p-1.

beyond the immediate front, above the hills, and beyond the forests is natural. Many writers notable among them like H.G. Wells created a myth and fantasy of air war where aircraft can wreck havoc from the air. Artillery's tactical doctrine was threatened in Boer War[10] when the gunners, guns and observers were forced to take cover because of Boers effective long range rifles, they reacted swiftly by employing the balloons of Royal Engineers to make sure their rounds are falling correctly at far distance yet the gunners retained the command of fire in their hands, the executive word of fire was the privilege of artillery commander and not that of observer unless he happens to be the same.

Since Montgolfier brother's hot air balloon ascent, the speculation of aerial war started brewing. The German domination in airship had aroused the public interest in the air defence of London fermented by the popular literature. In 1902 French Jurist Paul Fauchille had suggested freedom in air similar to freedom in sea.[11] The crossing of English Channel by Bleriot in 1909, made the heading of 'England no more an island'[12] by Daily Mail. In Britain there was a keen eagerness on part of the public headed by the air enthusiast, Member of Parliament, and influential people like Lord Montagau, who kept pressuring the government to take steps regarding future air threat. In 1911 a demonstration was held at London in which a mock ship was destroyed by dropping aerial bombs. First an air battalion was formed on 1st April 1911 at Farnborough with an establishment of 14 officers and 150 men finally on 13th April 1912 The Royal Flying Corps was raised. Applicant for aircrew generally entered the RFC as a cadet via the depot pool for basic training. The cadet would then generally pass on to the School of Military Aeronautics at either Reading or Oxford. Following this period of theoretical learning the cadet was posted to a Training Squadron,

10 Geoffrey Parker,ed;*The Cambridge History Of Warfare*,{NewYork:Cambridge University Press,2005},pp.256-259.

11 First AirWar,p-10.

12 Paris Michael ,Winged Warfare [,Manchester University Press,1992,] P 66.

either in the UK or overseas. Colonel Robert Smith-Barry appalled at the poor standard of newly trained pilots and high fatality rate during training in 1915-16, formulated a comprehensive curriculum for pilot training. The immediate effect was to have halve fatalities in training in 1917. The curriculum was based on a combination of classroom theory and dual flight instruction. After flying 10 to 20 hours dual instruction, the pupil would be ready to 'go solo'.

It were the aircraft of RFC which caught the imagination of the gunners and were used extensively for observation[13] and correction of fire, new gunnery procedures were evolved to facilitate the aircraft advantages in speed and limitation of endurance. One of the initial and most vital uses for RFC aircraft was the spotting of artillery fire. The results of the artillery fire were easy enough for the pilot to observe; the problem was communicating any necessary corrections to the firing battery. The early method was for the flier to write a note and drop it to the ground where it could be recovered. The RFC pioneered experiments with radio transmitters in their aircraft. Unfortunately the transmitters of the time weighed 75 pounds and filled an entire seat in the cockpit. this meant that the pilot had to fly the aircraft, navigate, observe the fall of the shells and transmit the results by Morse Code himself. Also, the radio in the aircraft could not receive so the pilots could not be sent any instructions or questions from the ground. This work was originally done by a special *Wireless Flight.*

Almost all belligerents had means to drop bomb from air either in form of aircraft or airships. France in 1914 had over 1000 Frenchmen with pilot's licenses and a year earlier Fredrick Janes had estimated that there were over 2000 aircraft in the World[14]. The general perception was that aircraft are primarily meant for aerial reconnaissance and artillery

13 First Air war, p-40.
14 First AirWar, p-12.

fire adjustment, however on 19th January 1915 England was bombed from Zeppelin, all in all thirty five tons of explosives were dropped from air in 1915 killing 185 people in England. Raids continued in 1916 and in 1917 as well. Germany also undertook an aerial relief mission involving a distance of over six thousand kilometres to support their forces in Africa.

Smut Report 1917

British parliament gradually accepted that air power can undertake the land forces task at a much more economical cost. It was Lord Montagu[15] who while addressing the House of Lords in 1916 stressed upon combining RFC and RNAS.

This cabinet report which was put up on 17th August 1917, is in fact the first ever concrete political analysis of the aerial warfare. It highlighted the wastage that is being incurred in maintaining two separate air wings for army and navy, more importantly it recommended an independent air force and a separate air ministry. Prophetically it stated ‘As far as can at present be forseen there is absolutely no limit to independent use of air and the day may not be far off when aerial operations with their devastation of enemy land and destruction of industrial and populous centres on a vast scale may become the principal operations of war to which all other forms of military and naval operations may become secondary and sub-ordinate’.[16] It was in April 1918 that the first independent air force Royal Air Force was established with Hugh Trenchard as its chief.

---

15 Winged Warfare ,P 82

16 Robertson Scot,Development of RAF Strategic Bombing Doctrine,{London,Praeger Studies} P17

Birth of Strategic Air Power

An airpower is strategic when it is capable of achieving the strategic goals of the war, when its employment can bring strategic dividends which in military terms simply means; bringing an enemy to accept our will.

The strategic concept of air power is derived from the independent air force of Trenchard in WW1, it had a simple notion that war can be won by destroying the people's will to fight, by destruction of means to wage a war. This was further cemented by the scientific calculation of Douhet's 'Command of Air' which stated how a whole city can be burned by using incendiary, explosives and biological bombs, there by making the civilian suffered and therefore through them bringing the war to an earlier end rather than the prolong stalemate of trench warfare.

The concept of strategic air power strengthened from the RAF birth and teething period when out of sheer necessity it came up with the idea and doctrine of winning the next war at its own.

Marshall Hugh Trenchard; 1873-1956

Trernchard was an infantry officer who had played polo at Sialkot with Winston Churchill, he was wounded in Boer War and joined RFC in 1912. RAF had over 22000 aircraft in 1918 and was reduced to mere 1000 in 1930. Lord Trenchard who remained RAF Chief between 1918-1928,states in his memorandum to chief of staff's in 1928. Aim of RAF[17] will be to breakdown the enemy means of armed resistances by attacks on objectives selected as most likely to achieve this end by targeting his factories, oil refineries, communication centre. The attack on enemy air-bases are not the main aims, air superiority will be gained through dog fights and not by direct destruction of an air

17 Trenchard, Memorandum to Chief of Staff committee ,2nd May 1928

force. Trenchard is thus favouring an independent air force which has no obligation to army and navy and neither advocates any integral air arm for these services.[18] He thus from British point of view advocates air force which can finish a war on continent without using the land forces and relying on destruction of enemy's will power, morale and fighting capability. Trenchard was the first chief of RAF, later he was removed from command, it was due to his old polo buddy Churchill who brought him back.

General Billy Mitchell; 1879-1936

Army acquired its first aircraft in 1909 America did not own a single aircraft when it entered the WW1 Which was even mounted with a flexible or synchronized machine guns[19], the strength of army pilots was not more than 70 with two squadrons, the Signal Corps had an aeronautical section established in 1907 that looked after the aviation it changed its nomenclature to Aviation section in 1914.[20] In September 1918 America adopted aprogramme of having over 200 squadrons. In May 1920 the American undertook the reorganization of their armed forces and resultantly the Air Corps was rais with a strength of 1514 officers and 16000 enlisted men it was third largest arm after infantry[4200 officers] and artillery [1900 officers] in terms of officers ratio 'there is here by created an air service the air service shall consist of one chief of air service with the rank of major general'[21].

An American Signal Corps officer who joined Air Service in 1917 and served in Europe and became the assistant chief of air staff between 1921-1925, before being court-martial. He is an extremist like

18 Gjert Lage Dyndal *Trenchard & Slessor on the supremacy of Air Power over Sea Power,Papers of the Royal Norwegian Air Force Academy, Volume 17*{Tapir akademisk forlag,2007}

19 Earl Meaden Captain, *Air service The General Services School,* {Fort Leavenworth,Kansas,1920},p-8.

20 Robert Finnley, *History of the Air Corps Tactical School 1920-1940*{USAF Historical Studies No.100,original imprint 1955},p-9.

21 Section 13a & 127 of National Defence Act approved on June 5th 1920 by US Congress.

Douhet and yet refrain from concept of Total War, rather he advocates 'It is not necessary to destroy complete cities, few bombs are enough'. His thought process is in line with Trenchard when he also states 'It is now realised that the hostile's main army in the field is a false objective, the real objectives are the vital centres'. Mitchell real efforts were directed to attain an independent status for air force in America, which did not realise till 1947. Mitchell was court martial in 1926 for raising his voice through the newspapers, today his statue is adoring the Udvar Hazy Smithsonian Museum at Washington. Billy was fundamental in bringing forward changes such as that America on 2nd July 1926 change the name of its Air Service to Air Corps, which wore their own uniform and had a promotion rate comparable to the other branches apart from having flying pay. American Army also had similar issues with American Navy over the control of air ,in the end the range of army's coastal artillery became the dividing line for army and navy's domain of air power. In 1931 MacArthur-Pratt agreement formally divided the responsibilities between these two services, another fundamental decision was that the command of tactical units (Air Corps) in the field should be in the hand of field commanders where as general reserve of attack, pursuit and bombardment to be concentrated under direct command of General Headquarters.

Mitchell advocates that Air service is the first line of national defence and it can react faster than the other two services 'an airforce is able to bring fire to bear in three dimension...able to concentrate on a point from three to fifty times more quickly than any other military element...by destroying enemy strategicalcentres,line of communication and reserves coming to the front,it either can bring about a decision'.[22] Billy also gets into the details of the organization of such air force,he highlights the importance and necessity of having air routes marked with navigational aids and aerodromes, his emphasis is on the defence of America against any invasion. Mitchell thus advocated the raising of

22 Lieutenant Colonel William Mitchel article, *The Mission of an airforce in the military organization of the United States* published in US Air Services, July 1920,Volume 3,No.6 {Air Services Publishers ,NY,}

independent air force at par with navy and army and not merely an arm equal in stature with infantry ,cavalry and artillery

Douhet Doctrine; 1869-1930

Guilio Douhet an Italian was born in 1869, he was commissioned in infantry and he in his capacity as staff officer was one of the pioneer Italian officer to comment about the aircraft and its potential in warfare. He commanded the aviation unit in 1911 and later commanded artillery as well. During WW1 he criticized Italian military incompetence and was court martial for a year however after the war it was overturned and Douhet was promoted as general of aviation. In 1921 he published his doctrine and theories in a book 'Command of the Air', Douhet died in 1930.

When Douhet wrote his doctrine 'Command of Air' in 1922, he was not the first one to write about the air warfare, but as is obvious from the preceding, it was there but not in a precise, concise, military and political manuscript, this is what Douhet accomplished. Douhet style of arguing is similar to Clausewitz way of writing.

Douhet[23] starts with the effect that in any future war basing upon the experience of WW1,it is going to be a total war and thus there should be no difference between the civil population and military as far as suffering is concerned, the military can stand the hardships but civil population cannot ,thus it should be the civil population, the cities which should be targeted and destroyed in a manner by using explosives, incendiary and gas bombs that ,their suffering should compel them to seek peace

The quickest and the [24]most economical way to do so is by employing aircraft, as they can penetrate into the heart of every enemy

23 Douhet Guillio,Command of Air, P18
24 Ibid,P29

city, more so Douhet has always kept the financial aspect and the reality of Italian state in mind while elaborating his doctrine. He states that for the cost of one dreadnought one can have over one hundred aircraft, which will have more firepower than the ship, thus aircraft is the logical choice for causing destruction. He selects a zone of fifty square meters and states that it can be destroyed with one aircraft and ten aircraft are required to destroy five hundred square meters area and fifty such targets should be destroyed daily by using five hundred aircraft and soon the whole country will be engulfed in utter destruction.

Douhet logically[25] admits that enemy can also employ the same method, thus what is the remedy, and he proposes that enemy air force should be destroyed on ground there by achieving command of air, and anyone who will have the command of air will be able to achieve the victory and any nation who does not have command of air will lost the war, even if its army or navy are intact because the home front will collapse, thus the first target of own air force will be enemy air force on ground.

To achieve this he advocates formation of an independent airforce, at par with army and navy. He pleads state to allocate separate budget for this and accord it the highest priority. Douhet also floats the idea of integral Auxiliary air forces for army and navy, who should remain under command to them and should be raised from their budget where as the Independent air force should only deal with targets to achieve command of air.

Douhet then elaborates what should be the composition of this independent air force, and favours bombers, but supported by fighters. Douhet technically admits that it is not possible to have an ideal aircraft for all missions and instead of a compromise thus favours specialise aircraft.

Douhet, stresses the importance of air as a commercial tool, highlighting that Italy should control the commercial airways by

25 Ibid P 31

encouraging the private airlines, he very prophetically states 'soon all international mail will be an airmail'.[26] Douhet debates at length to highlight the advantage which a civil aviation can give to military aviation in case of war. He thus goes to state that pilot training, technician training should be done by civil aviation to reduce expenditures.

Adolf Hitler; 1889-1945

Probably no man after Douhet has defined and practiced aerial warfare than Hitler, for the reason that Douhet advocates the aerial concept of Clausewitz's idea of 'Total War' and Hitler was the only military leader who embraced this concept. It was he who raised Luftwaffe in 1935, an air force which was unique in its organisation and mission. Like RAF it was independent, yet it worked in harmony with the army and navy. It procured aircraft which were tailor-made for the whole concept of 'Lightning War' and with the aim of war itself. Hitler incorporated the concept of airborne troops as part of air force, thus air force was not meant only to destroy the cities or support the army, it was to conduct independent operations where it will not only have the command of air but will also hold the ground by itself, it had integral troops, two airborne divisions. It was this unique concept of third dimension which captured Narvik, a strategic objective despite British naval superiority, these troops later captured strategic Eben Emael fortification through glider operations, it was Luftwaffe which through Douhet concept of destroying city, burnt Rotterdam forcing Dutch to surrender and finally the capture of Crete and morale boosting rescue of Mussolini. General Wever started an Ural Bomber project in 1935-36 to built a long range bomber to attack Soviet Union in depth, his death in an air crash in 1936 resulted in the abandonment of the programme and

26 Ibid P77

Luftwaffe thus adopted a medium range bomber force programme.

In between the wars the gulf between army and RAF grew to an extent where army did not had a single aircraft under its direct command.

The concept of Gunners or for that matter British Army having control of aircraft and aerial fire control over artillery finally matured in 1941 with the raising of Air Observation Post{AOP} Squadron, before VE a dozen such squadrons were raised .These squadrons were unique as they were jointly operated by the Artillery and RAF; not surprisingly both services had different opinion about their performance and utility in the war.

The existing written material which is limited to half a dozen books on the subject of AOP are almost entirely written by AOP officers. The very first official history was published in 1956 by the AOP Association and its forward was written by the Field Marshall Alan Brooke the war time Chief Of Imperial General Staff ' my feelings on the Air O.P. were unfortunately not shared by some of those involved in the original discussions'.[27] The theme set by the very first book revolved around that 'It is fair to level against RAF certain criticism…they think that theirs was the air and everything that was in it'.[28] Same theme has been repeatedly adopted by almost all historians writing on the subject including AO.P. pilots memoirs 'The Royal Air Force said it was impossible but the Air OP proved them wrong'[29]. This over emphasizing and zealot display of espirit-de- corps is also displayed none other but General Sir Martin Farndale who wrote the forward of another Air OP history book 'Their flying hours broke all records'.[30] On the other hand

27 Parham and Belfield,*Unarmed into Battle The Story Of Air Observation Post*{Warren & Sons,Winchester,1956} pix

28 Ibid p 18-19 and 21

29 Andrew,Lyell, *Memoirs Of An Air Observation Officer*{Picton,Wiltshire,1985},this sentence is written boldly on the back cover.

30 Jones,Maslen, *Fire By Order ,Recollection of Service With 656 Air Observation Post Squadron in Burma* {Red Wood Books,Wiltshire,1997}.

in the sphere of RAF history there is hardly ever more than a paragraph for the Air OP and in majority of field formations published histories the AOP at times is not even included in the order of the battle. Thus there seems to be a contradiction in the raising, history and operational evaluation of the Air OP. Every British regiment needs a hero and Lieutenant Colonel Charles Bazeley 'Father of AOP'[31] is the hero and RAF in general and Air Marshall Barrett in particular is the villain.[32]

While working on the dissertation an extensive access to the primary sources was available at National Archives Kew, Museum of Army Flying{MAF}Middle Wallop and Museum of Royal Artillery {Fire Power} Woolwich. With the evidence of these primary sources the existing conception and perception about AOP have been debated and particularly through its employment and war record in the various campaigns and battles following gospels of their history has been analysed.

First Gospel. Air Observation through Artillery officers flying and controlling the fire was more useful than the RAF pilots doing so.

Second Gospel. RAF deliberately delayed and created hurdles in the maturity of this concept.

Third Gospel. Lieutenant Colonel Charles Bazeley and Major General Parham are the prophets of this concept.

Fourth Gospel. These Air Observation Squadrons proved their worth in the WW2.

---

31 John Terraine,The Right Of The Line,{London:Sceptre,1988}pp-398.

32 Ibid,p.399

Chapter One

Artillery-Air Co-operation 1900-1939

British Artillery traces its history to 1716 when through a royal charter the Royal Regiment of Artillery was raised,since initially it was under the master general of ordnance ,this is the reason that artillery wears blue dress as compare to the scarlet of infantry and cavalry. British used their artillery in novel way during the Crimean war when they employed heavy calibre guns and laid rails for its logistic support. It was again the French who in 1890 developed the quick firing mechanism for their 75 millimetres gun, the remarkable feature was that it had a system where the recoil was absorbed by the gun carriage employing the hydraulic fluid.[33] The main draw back till this point was how to engage the enemy which is not visible to the gunners and neither the guns could be placed in front along with the infantry and cavalry for the reason that they were the prime target of enemy,[34] it was a Russian who invented the target acquisition and engaging system by employing various signals of flag and utilising the field telephone in 1890.

Boer War (1899-1901) was the first conflict in which British employ this tactics of having guns in rear and correcting their fire through an observer yet the limitation of how far the observer will be from the guns and how far he himself can see the fall of shot were the primary hindrance in having an effective fire. The desire of every commander was to have a direct hit with each round of his guns the way

33 Firepower,pp-9-27.

34 Shelford Bidwell Gunners at War a Tactical Study of the Royal Artillery in the 20th Century{Arms & Armour Press,London,1970}.pp,15-16.

an infantryman thinks of killing foe with each round he fires. In terms of guns the inherent limitation is the varying of charge in each shell the meteorological changes and wear and tear of the barrel itself, thus it is virtually impossible to have two rounds falling on exact spot.

The tactics of employing guns remained virtually unchanged in over two centuries till Boer War when it became almost suicidal to deploy guns along with or in front of infantry or cavalry. Thus there was a need to deploy the guns in relative safety yet the requirement of getting its shells on the enemy unseen to the gunners remained same. In the initial days of Boer War the battery commander used to pass the orders to the guns standing close to them and observing the fall of shot as well using megaphone later he would use flag signals or line telephone for the same purpose. Artillery fire is thus classified in terms of direct fire when gunners can actually see the target and fall of shot and apply corrections to it or indirect fire when an observer passes the correction to them. But how can an observer see the target which is behind the hill or in jungle or beyond the visual range of the eye, the solution and natural answer was only if he can get in air or be enough forward and yet he should have the communication means to pass the orders thus concept of forward observer officer was evolved in Boer War and it was also in this war that British army employed the balloons of Royal Engineers for correcting artillery fire for the first time.

Three balloon sections arrived at Durban on 26th October 1899 and were employed at Magersfontein on 11th December 1899. In April 1900 balloons were used for directing the fire of 5" howitzer of 37th Battery[35] at Megersfontein when 65th battery were enabled by the balloon observers to search with fire. Communication between the balloon and guns was in infancy; mainly through shouting and using line telephone. It

---

35 Peter Mead,The Eyes in the Air,history of air observation & reconnaissance for the army 1785-1945{London,Her Majesty's stationery office,1983},p-25.

is worth mentioning that signallers in Royal Artillery were abolished in the batteries in 1899 because it was felt that line telephone presence does not warrant them, previously these signallers were employed to convert the fall of round and related corrections as ordered by the battery commander, these signallers were reintroduced in artillery during Boer War.

Royal Flying Corps was formed in 1912 from the nucleus of the same balloon sections which in 1908 were first raised to the status of air company later into air battalion still being integral to Royal Engineers. In August 1912 War Office held a military competition for the selection of an appropriate aircraft, whose salient features were desired to be following.

Pay load. 350 pounds
Endurance. 3 hours
Take off with above pay load from long grass, clover or harrowed field within 100 yards and should be able to land without any damage within 75 yards.
Dual controlled
Good field of vision
Engine able to be started by the pilot alone
All parts should be interchangeable
{In the end the closet to come to these requirements was Royal Aircraft Factory BE2}

The RFC's first fatal crash was on 5 July 1912 near Stonehenge on Salisbury Plain. Killed were Captain Eustace B. Loraine and his observer, Staff-Sergeant R.H.V. Wilson. An order was issued after the crash stating "Flying will continue this evening as usual", thus beginning a tradition

The 1912 field regulation specified that 'cavalry and air service were complimentry to one another'.[36] Between 1912 and 1913 there were

36 Eyes in the Air,p-44

efforts made to establish an "artillery cooperation role " for the RFC and No 3 Squadron was made responsible for this and trials were carried out during 1913 army manoeuvres in Scotland ; major limitation was the lack of communication between aircraft and the ground. Another aspect was the lack of trained artillery observer in the aircraft[37] thus the very first observer course started in July 1914 and a procedure was agreed upon between the RFC and Royal Artillery for communicating fall of shot.

With these primitive tactical and technical thoughts of air co-operation the British Expeditionary Force{BEF} arrived in France under General John French with four divisions of army and almost four squadrons of RFC whose primary and some what sole role was gathering of information. RFC No.2,3,5 squadrons were concentrated at Amiens on 13th August 1914, two days later the No.4 Squadron also joined in,they moved to Marbeuge, on 19th August 1914 the very first reconnaissance sortie of the war was flown by RFC,on 13th September the very first artillery cooperation sortie took place at Aisne and on the same day the very first photographic sortie also took place here. By late 1914 the BEF was organized into two separate armies thus RFC was also reorganized into wings,one for each army and each wing had four squadrons there by having one squadron each for a corps.[38] On 26th August 1914 the first major battle between the BEF and Germans took place and Royal Artillery learnt a bitter lesson in the battle of Le Cateau when the guns were deployed in the front and in open resultantly they suffered heavily for this, it was probably the last time that guns were ever deployed so close to the infantry and in open, 'artillery moved to the rear for the safety'.[39] It had an adverse impact on the relation ship between the three arms{infantry, cavalry, artillery} guns from being a direct firing weapon became an indirect support thus regiment lost its face to the other two

37 Eyes in the air,p-69.

38 Shelford and Dominick *Fire Power*, p. 68; Ernest Swinton, ed,*Twenty Years After, The Battlefield of the 1914-1918,Then and Now volume1* , {London: George Newnes Limited,1939} pp.63-68.

39 Ibid

in any case all three arms infantry,cavalry and artillery used to dined at different tables[40] .Artillery reliance on RFC now became almost compulsory to complement the accuracy of fire and to reconnaissance for targets, this task was thus called artillery reconnaissance {Arty R} and other army reconnaissance became tactical reconnaissance {Tac R}. Arty R was some what different from artillery observation as in the former the pilot would remain in air and correct the fire of the guns.

Before the end of 1914 RFC was organised into wings and squadrons for the corps, yet there was only one squadron {No 9}which was equipped with the wireless for artillery observation. By end of 1915 each corps had its own squadron equipped for tactical and artillery reconnaissance. At that time the artillery observation from air was very slow and tedious task, there were no two way communication between the aircraft and the guns, in some aircraft there were wireless sets which were very heavy and it was not easy to pick up the Morse Code amidst the noise of the engine. There were thus trained artillery observers in the aircraft that would carry out these tasks and observations would be dropped through message writing from the aircraft or aircraft would land beside the artillery battery[41]. In 1915 the War Office issued a pamphlet 'Co-Operation of Aero planes With Artillery'.[42] It dealt with the employment of aircraft in support of artillery. War Office highlighted the necessity of close co-operation between artillery and RFC headquarters at all times stating that aircraft can assist the artillery which is supporting the infantry immediately in front by locating the targets and giving approximate range '*it will be of great advantage if the observer is able to bring fire to bear immediately*'.[43]

40 Ibid,p-3

41 Lee Kennett,*The First Air War 1914-1918*{Toronto:the Free Press,1991},p.33

42 WO 158/681; RFC Co-operation with Artillery letter from CIGS to G.O.C- I.E.F. 'D' Basradated 1915,see also WO 158/682,683 and 684 dealing with general questions about RFC.. H.Wilson and J.Hammerton, ,ed,*The Great War The Standard History of The All-Europe Conflict Volume 1,A new arm in the element by an expert in aviation*{London:Amalgamated Press Limited ,.1914}pp-315-346:

43 Ibid,p.3.

Flying in figure eight over the battery signifies that the target being ranged upon is firing or is on the move. Three red lights mean that the observer considers it necessary to stop the fire owing to the possibility of danger to own troops. Stop signal followed by flying in figure eight means that the observer has discovered a fleeting target of superior tactical importance, the aircraft will then proceed vertically over the target and will fire a smoke ball at 6,000 feet this gives the necessary data for opening round.

Another technique in which observer in an effort to designate new target used signal 'lay on me' the aircraft fly along the line of fire, fires a prearranged visual signal and reports the approximate range to the objective, the ground artillery was thus able to calculate the height and range and thus getting the two sides of a triangle.[44]

Message	**Morse Code**
Are you receiving my signals	B
Stand By	A
Fire	G
Over	O
Short	S
Right	R
Left	Q
Washout	WW
Am returning to ground	C.B.

The point to note is that War Office admitted, expected and promoted the fact that pilot should engage the target as quickly as possible and should not follow the deliberate ground procedures which in any case were dependent upon the individual battery commanders, some of them would give the correction of the day basing upon the drift

44 Ibid

of their morning cigar's smoke.

An artillery gun or combination of guns{battery/regiment} can achieve at best while firing indirectly is an area destruction, since target is not visible thus the battery commander relies upon either map or the ground observation which can take over fifteen to twenty minutes just to hit a target yet fire is not guaranteed to fall at the same spot because variables like wind and wear/tear of gun barrel differs daily. This luxury of time was not available to the pilot as fuel endurance was the key factor thus observation had to be quick and immediate. Royal Engineers Lieutenants D.S. Lewis and B.T.James along with Major W.Salmond of RFC devised a simple **'Clock Code'**[45] method of indicating target to the battery and its immediate neutralisation, assisted by the gridded mapping of the Western Front with maps of 1/20,000. It was simple in nature, target was imagined to be the centre of a clock with its axis running true north and south, 6'clock being the southern point, various concentric circles at radial distances of 50,100,200,400 and 500 yards from the target lettered 'A' to 'F' enabled the fall of the round achieving an accuracy to the tune of fifteen–twenty yards. Royal Artillery apart from the aircraft also employed balloons and kites for target observation and correction. Balloons were more effective as they could remain in air for a long duration but they were flying ducks for the German fighter aircraft. In 1915 inventor Everard Calthrop offered the RFC his patented parachute. On 13 January 1917, Captain Clive Collett made the first British military parachute jump from a heavier-than-air craft. The jump, from 600 feet, was successful but the higher authorities in the RFC and the Air Board were opposed to the issuing of parachutes to pilots. It was felt at the time that a parachute might tempt a pilot to abandon his aircraft in an emergency rather than continuing the fight.

---

45 Bidwell, ,and Graham,*FirePower* p.105.,Mead, *Eyes in the Air* p. 103.. *Twenty Years'More about the Guns'* pp. 219-228 and '*With R.F.C.In The Great War* pp.484-499.

The parachutes of the time were also heavy and cumbersome, and the added weight was frowned upon by experienced pilots as it adversely affected aircraft with already marginal performance. It was not until 16 September 1918 that the order was issued for all single seat aircraft to be fitted with parachutes the observers in balloons were also supplied with parachutes. One balloon observer wrote 'I could see the enemy till 10,000 yards....artillery officers generally did not like this observation from balloon'.[46]

Wireless Telegraphy. It came at the end of 1914 ,it was cranky and pilots were given set of block time to transmit, this was later improved and observer could tap out their message in Morse Code. germans improved wireless reception by end 1917. The RFC aircraft would reel an antenna of 100 feet long, ground receiver was in the battery, the ground signals were often smoke bombs, cloth panels and signal lamps. The Italians used the cloth signals in 1917 to indicate 'yes' 'no' 'battery ready' 'land' 'am suspending fire' & 'I do not understand'.[47]

Photography. In early days in RFC the observer used to take the photographs from their own camera, in February 1915 the A Camera were introduced, by the end of war pictures from 15,000 feet were taken.[48]

Royal Regiment of Artillery was bit unusual in the other wise eccentric British Army as they preferred to call their combination of two batteries unit commanded by a lieutenant colonel as 'Artillery Brigade'. The command structure of artillery was also different, thus an artillery brigadier does not have any real command, he is advisor to infantry or cavalry divisional commander and all his orders had to pass through the 'G' staff of the division, at higher level again an artillery major general

46 www.imperialwarmuseum .
47 First Air War, p-34.
48 Ibid p-36.

is merely an advisor to the corps or army commander, therefore artillery developed its own psyche in which despite their fire power and sheer size of force they were not a fighting arm but a supporting one. Thus it was difficult for the artillery commanders at all level to accept the fire orders from an aerial observer it was taking away the last vestige of honour from them, thus they resisted direct orders but accepted and expected the observation.

After the armistice both Army and Royal Navy demanded the return of their aircraft wings and there was a feeling to dissolve RAF, but it was due to Major General Trenchard that RAF survived on the plea of empire policing and motto of 'army policy is to defeat the enemy army ours is to defeat the enemy nation'. A series of reports and inquires started in 1923 the Marquess of Salisbury inquiry on 9th March 1923 was to inquire into the co-operation and correlation between navy and RAF and corresponding relationship between army and RAF,it rejected the War Office demands of handing over the Air Ministry to them War Office had raised the objection about extravagant life style of RAF and duplication on dental, medical and chaplain expenditures. Royal Navy under Admiral Beatty were adamant to have the flying wing being resorted back to them purely on grounds of operational requirement, as Beatty said 'Air arm of the navy is a naval unit of the fighting fleet exactly similar to the light cruiser or destroyer arm'.[49] Army on the other hand did not took air seriously, neither it pursued as vigorously as Royal Navy, did rather it attack the 'basic moral fibre of RAF officers'[50] Lord Derby remarked 'calibre of young officers {pilots in RAF} who are taken in is very low' and even proposed that every army officer in future would be trained in flying the way they are being trained on riding 'you could

49 Hyde, Montgomery, *British Air Power Between The Wars 1918-1939*{London:Heineman,1976} ,p.109.

50 Ibid,p.141.

have the best men,if all those who joined army undergo a two year flying course and then if he is not found fit can go back to his unit'. RAF thus gave away to certain concessions more to RN, Trenchard-Keyes agreement of 1923 made it possible that almost 70% of Royal Navy air observation reconnaissance gunnery duties will be performed by the pilots selected from the Royal Navy, the Balfour report of 3rd August 1923 recommended that ratio of naval officers seconded to RAF be increased from 20% to 30% and raising of Coastal Command. Trenchard conceded less to army that included having officers from both services on secondment yet a lasting friction developed among RAF and army based upon personal egos. Artillery which emerged out of the Great War as the most accomplished arm in terms of number of troops and promotions started fading out not only to RAF but also to tanks. In 1929 a leading artillery officer contemplating about future debated upon the idea of artillery commander controlling the fire from air but then he rejected the idea in next sentence as it would be contrary to customs of service for an artillery commander to be away once the divisional commander is in office. He also negated the thought of having a gunner as pilot as in his opinion it is more easy to teach a pilot of RAF about gunnery than to teach a gunner about flying.[51]

Artillery in the post armistice gain a lot of credibility and offset the traditional hierarchy of the organisation which was primarily infantry or cavalry oriented. The number of gunner generals out numbered both infantry and cavalry, artillery had 34.7 (26 in number) as compared to 4% of cavalry and 61.3% (46 in numbers) of infantry. The lieutenant generals of artillery were 43.5% (10 in number) as compared to 56.5% (13 in numbers) of infantry; cavalry had none. CIGS from 1923-1932 was a gunner; Field Marshall Milne followed by another gunner General Sir Montgomery-Massingberd however there was no great interest or

51 Robinson Rowan Colonel, *Artillery Today and Tomorrow*,{London,William Clowes &Sons,1930},pp.52-60.

enthusiasm for things mechanical.[52] By 1927 all medium artillery was tractor drawn ,by 1937 nine field brigades were mechanized.The training memorandum of artillery in 1939 stressed 'The type of war we must now consider is not the highly mobile type but a more ponderous type of masses'.[53]

First Great War was a military revolution; General John French led a British expeditionary force of seven divisions whose heavy artillery consisted of 24 five inch guns by 1918 General Haig had over 60 divisions supported by six and half thousand field pieces from three to 18 inches in caliber.[54] The use of third dimension had a profound affect on all the existing ground weapons ,more so on the artillery which was and is still a long range weapon, where human observation is unable to match its long range to pin point the target and to accurately direct the shots. Thus Royal Flying Corps {RFC} and Royal Artillery almost became an integral part to each other. By the end of the war the Royal Air Force {RAF}found itself capable of achieving by itself the same tasks for which it was subsidiary to artillery, yet artillery was left high and dry as by then it was relying upon aircraft, balloons and kites for effective gun fire. After WW1 the RAF formed a army cooperation group to provide the army with all kind of reconnaissance and few years later agreed to have army officers as pilots on attachment secondment wearing RAF uniform. The inertia of British Army in accepting changes is obvious from the fact that in the spring of 1937 there was a rumour that one of the Scottish regiment 'Scot Grey' is to be mechanized,in response its commander wrote to leading Scotts and resultantly the War Office was flooded with letters not to mechanized it or to make it the last one.[55]

---

52 Philip Venlar & David fletcher *Moving the Guns the Mechanisation of Royal Artillery 1854-1939*{London HMSO August 1989}.p-23.

53 C.A.L.Grahaam, Brigadier General. Story of Royal Regiment of Artillery{Royal Artillery Institution,Woolwich,1962},p-70

54 Ibid,p-1.

55 British Army & the Theory of Armoured Warfare 1918-1940.p-28.

Gyro was invented by a Spanish Cerva in 1926 and it had a remarkable ability to operate from rough fields to land back in a confined area. Captain H.J. Parham, in an article in 1933 remarked that 'The battery commander, wishing to see a target out of sight behind a hill, is compelled to call upon a unit operating from a distant landing ground when five minutes in the air above his own battery position...even a thousand feet up...would meet all his needs'.[56] Royal Artillery Flying Club (RAFC) was formed in November 1934, the club did not own an aircraft rather it extended financial help to the gunners to learn flying; Brigadier Massy was its first president. Under the pressure of gunners in 1935 the first trials of 'Rota' an auto gyro were carried out mainly for the Royal Artillery.[57] It was flown by two RAF pilots with two gunner officers Lieutenant Mead and Major Norman Langs on board to carry out the trials of observing aerial observation, 'we took half a dozen shoots…my battery commander could not observe the fall of the rounds and felt that flying control should be in his hand'.[58] Rota was an ideal aerial platform for artillery tasks as it could fly very slow and it was difficult for any aircraft to locate it during flying. Later these Rotas were not put under any trial due to technical reasons. There were quite a few artillery officers who have been flying Rota, in fact the very first gunner officer who is credited with the idea of floating air observation post, Major Parham[59] {Later major general}owned one at that time.

In early thirties the preparation for a future war started in earnest with RAF taking the gauntlet and emerging as the most viable saviour, relying heavily upon its bomber force to an extent that it over looked the fighters almost totally. The question of army-air co-operation was on low priority yet a new multi purpose aircraft Lysander was in making, designed to meet almost all of the army requirements

56 Parham and Belfield, *p-10.*

57 WO 32/2840 outlining modernisation of arty till 1937-1938..

58 Peter Mead, „*Soldiers In The Air, The Development of Army Flying*{London:Ian Allen,London,1967},pp.5-9.

59 Parham and Belfield, *Unarmed into Battle* , p.10.

that included tactical and artillery reconnaissance provision of aerial photography was another task of these aircraft. RAF was not totally insensitive to army needs but the time, resources and over all strategic posture that included 'Knock Out Blow'[60] created an atmosphere of almost mistrust in the army hierarchy and it was not helped by the fact that RAF increased the number of aircraft for army support only from 90 to 120 in its expansion programme. On the other hand army itself was in a state of inertia, in the first ever review of first great war 'Kirkee Report 1931'[61] every thing was fine with the army and artillery despite the fact that artillery was still bound to horse.In the same time period Royal Navy{RN}was still fighting for its control of the Fleet Air Arm{FAA} with RAF and they finally won the battle in July 1937 under 'Inksip Report'.[62] And that was the catalyst for the Khakis to also have a share of air cover and none other than the Royal Artillery, the most technical, educated and yet not treated at par socially in army customs came in fore front for revival of air co-operation of by gone days by exploiting the Clock Code method. Surprisingly army never considered having its own air wing merely for casualty evacuation as part of Royal Army Medical Corps{RAMC}.

Captain Charles Bazeley{Lieutenant Colonel retired}remained Royal Artillery Flying Club secretary in 1934 and learnt his flying there, Bazeley in 1937 was seconded to RAF. He is regarded by almost all the published material as the founding father of Air Observation Post{AOP}[63] by virtue of his 1938 for the annual Duncan Artillery Prize Essay .It is important to analyse his thoughts in detail, first he did not won the essay it was won by the officer who wrote that present day RAF procedures are good enough. Bazeley had limited flying experience and his thoughts reflect that.

60 Jhon Terrence Right Of Line
61 Fire Power p 150,187,.see also WO 32/2840
62 British Air Policy between wars,p 402,.fire power 261-263
63 His picture has the caption of Father of Air OP at Museum of Air Flying, Army Air Corps School Middle Wallop England.

'Aircraft is an ideal observation post...on 90 days out of 100 when it is possible to fly it is possible to rise to 1000 feet ,on 7 out of 10 days it is possible to see 6-7 miles horizontally...army co-operation pilot has to undergo a highly technical training, he has to send message by Morse Code and receive them in air... in short he{RAF pilot}is a superman...nor it is necessary that the gunner's flying training should be comprehensive...modern aircraft requires very little maintenance... Observation constitute a major part of all artillery ...control is exercised by a pilot who cannot be trained as a professional gunner...pilot cannot select targets on ground as he is not a gunner, pilot can control only one troop at a time...aircraft should be provided on a scale of three per field regiment and medium regiment.'[64]

He also highlighted the difficulties which are being faced by the artillery in aerial observation as the pilots are RAF pilots and who does not speak gunner's language. Bazeley's idea was that an aircraft should be placed next to every battery and battery commander should simply get airborne after just twelve weeks of flying course.

There was nothing wrong with the existing system as far as the tactical environments were concerned, an aircraft remains an aircraft irrespective of whom is flying, a RAF aircraft can pick up a target at low level and can pass the data to the ground, provided it has got the communication frequency, RAF aircraft can also knock out opportunity targets as well, they can be flown by the artillery officers or vice versa the pilots themselves learn the artillery fire procedures which in any case is more economical and easy than teaching gunners how to fly but these aircraft will remain under command of RAF that was the air force point of view. Army on the other hand in its quest for aircraft to be under its command simply overlooked the tactical limitations of artillery aircraft and refused to budge. It was thus in 1938 that artillery made an effort

---

64 Royal Artillery Journal October 1939,pp 313-334,see also Unarmed into battle p 115,eyes in air pp 56-64,soldiers in air pp,memors of air op pp 98,fire by order pp76,right of line 443,

to have the aircraft for the same tasks of WW1 by objecting to the existing system and terming it as slow, unreliable and ineffective. It is in this background that the famous letter was written which set the scenario for coming years.

On 12th April 1938 the General Officer Commanding Southern Command wrote a letter to Under Secretary of State at the War Office in which he highlighted the importance of the artillery and how its lethality is getting impotent due adherence to First Great War's techniques and procedures of observing and directing the artillery fire from air, General educated War Office on the existing system being employed by the Royal Air Force{RAF}pilots in which they are using 'Clock Code' method for correcting the fall of shell .

'Present system is a relic of the war of 1914-1918 and but little advance has been made…this system has following grave objections… concentrated fire cannot be controlled from air …aircraft are compelled to fly at low heights over the enemy lines where they will be subjected to an ever increasing volume and accuracy of fire…{RAF}pilot is unaware of tactical situation'.[65]

General as a remedy suggested to War Office that if the pilot is a gunner officer and take off from the ground close to the artillery battery then the situation can improve, he also informed the higher authorities that already there are over hundred gunner officers who holds private flying licences and they can be employed in this venture, in the last he strongly recommended that only limitation in his other wise undisputable idea is that 'observation from behind our own lines may not be possible to a sufficient depth'.[66] War Office after two months send the almost same contents to Air Ministry for their comments.[67]

65 WO 216/77 ; S.C.No 5/1458{R.A.} ; Notes on the History of FOP,also MAF Document No 11.
66 Ibid, p-2.
67 Ibid, War Office Letter No.43/artillery/456/M.T.29th June 1938.

The basis of this letter were not the results of any army manoeuvres or trials nor any army study on the procedures in vogue rather it was the conviction of Brigadier Royal Artillery {BRA} the senior artillery advisor in the southern command Brigadier Massy who was obsessed with the idea{as later events proved .

Air Ministry 's reaction to the letter of southern command on 23rd July 1938 was professional.. They highlighted that in WW1 same procedure was used with effectiveness and aircraft does not have to fly low to observe the fall of rounds rather observation in that war was conducted from an elevation of ten thousand feet and above. Air Ministry further pointed out that under present tense international situation it would be much better if the existing system is improved rather than venturing into a new one as the 'system has been the basis of all post war training'.[68] War Office's rejection of these logics was reminiscences of verbal bouts these two offices had in early twenties and they reflects the military culture which finally led to the disaster of Dunkirk .War Office insisted upon carrying out trials in June 1938 to which Air Ministry reluctantly agreed[69] in its 23rd December 1938 letter, it was also agreed upon that three artillery officers with varying experience should carry out the trials. No 22 Army Co-operation Group RAF and School of Artillery conducted these trials at Larkhill in February 1939. Artillery officers led by Captain Bazeley flew Taylorcraft aircraft which was owned by the RAFC and evaluated it against the RAF army support aircraft Lysander. At shorter range the slow moving aircraft was quick to adjust the fire and also pick the fall of round more quickly than fast moving Lysander but at range beyond 5000 yards it was Lysander which was effective and Bazeley himself admitted this that 'they are unable to see the fall of round at

---

68 WO 216/77 Notes on the History of the FOP now called AOP,Air Ministry reply 768959/38/S.6

69 AIR Report on Army/Air requirement ,pp.144-145.

that distance on a clear day'. [70]

The result of trials became controversial as both services having varying concepts. Army or rather artillery pressed home the advantage of low speed aircraft as it saves time and RAF highlighting the range limitation of the aircraft. Army had no in depth air knowledge other than the RAFC officers who more as a regimental pride were keen to have the aircraft on the other side. RAF senior officers were practically all army officers who had taken active part in the WW1 as observers or had flown these artillery related missions. For the rest of the year the trials went on using four different types of aircraft including Cierva Gyro[71] but it was rejected by Bazeley on grounds of maintenance problem RAF was mentally occupied in a more strategic game than to be dragged into such affair and were right in pointing out that its more appropriate to mend the existing system because time is the key factor and precisely this is what happened, from 1935 onwards Germany was perceived as the real threat and therefore the bulk of RAF strength was supposed to be in bombers. The first four engines bomber 'Short Stirling' joined service in August 1940 and other medium bombers 'Avro Manchester' and 'Halifax Handely Page' joined in November 1940. thus at the beginning of war none of RAF bombers were capable of reaching Berlin 900 kilometers away .One important weapon to fight this menace was Radar, which in 1939 was able to detect approaching aircraft at a range of 100 miles from the coast, Britain had twenty such radar stations.[72] In May 1936 important structural changes took place in RAF, dividing it into five sectors corresponding to geographical areas in England, it also had five overseas commands as well. It had Bomber, Fighter and Coastal commands. In the same year RAF was authorised to increase its airfields from fifty-two to one hundred and twenty eight.

70 CAB 65/2 8th November 1939
71 D Flight papers,FOP trials,extract from No 22 Group Orb dated January 1940,MAF folder No.5
72 Ibid,p 77

In 1938 a balloon command was also raised for air defence barrages around cities, in 1939 the Fleet Air Arm was returned to admiralty.

Control of British Army is vested with Army Council, secretary of state for war is the president of the Army Council and responsible to the king, cabinet and parliament for all affairs of Army Council, certain responsibilities in regard to the element of British Army in India are however assured by Government of India. The military members of the Army Council were the Chief of Imperial General Staff (CIGS), Adjutant General, Quarter Master General and Vice Chief of Imperial General Staff (VCIGS). The land forces of UK comprised of regular army, territorial army and the reserve forces. On first September 1939 the strenght of regular army was 224, 188, Militia, 34, 542, Total 2,58,730. Army reserves were 1,40,500. Out of this strenght nearly half were abroad, a quarter in India. Service in regular army was 12 years.

Army in 1939 had five infantry divisions each having 12 battalions, one cavalry/armour division. In 1927 Imperial Defence College was established. British Navy is governed by Board of Admirality, the First Lord of Admirality is the Cabinet Minister responsible for Navy.

Commander in Chief Home Forces was a post created in 1940, under him the country was divided into five commands, in pre-war the commands were directly responsible to War Office.

Chapten Two 1939-1941

Flying Observation Post

to

Air Observation Post

On 1st September Germany invaded Poland and on 3rd September 1939 Great Britain declared war on Germany. At the out break of war in September 1939,[73] RAF had 1,476 aircraft at home and another 435 aircraft were deployed abroad, it had a manpower strength of 118,000 with another 68,000 as reservists. On the out break of war Luftwaffe possessed over 3600 front line aircraft with a manpower strength of half a million with another half a million in air defence which was also under Luftwaffe.The most distinct feature of Luftwaffe were its airborne troops. RAF had deployed ten squadrons out of its thirty-three bomber squadrons in France, these squadrons arrived in France on 2nd September 1939.[74]

September 1939- May 1940

Britain and France were at war with Germany, thus it is surprising that in this period nothing significant was done by them to stem this flow.[75] 'The war cabinet had realised that bomber command is incapable of delivering a knock-out blow…due to fear of German reprisal no

---

73 Ibid p 102
74 Ibid p 102
75 Ibid p 102

bombing was conducted…RAF was task to attack the war ships and drop leaflet'.[76] The first British air mission was on 4th September 1939 in which fifteen Blenheim and fourteen Wellington bombers attacked the German naval bases in and around Wilhelmshaven resultantly only one hit was attained at the cost of seven bombers lost, similarly by the end of December 1939 three major air attacks were conducted without any strategic success but with an alarming loss of 39% of force. French air force was much stronger than Luftwaffe in terms of bombers yet it virtually remained idle. This was the first phase of the war in which Germany was able to roll over Poland, Belgium, Holland, Norway and finally France. It looks an impressive tally especially when it is kept in mind that Germany was a late starter in the arms race, the major reason for their success was in the concept of the modern war, where Luftwaffe gained the air supremacy and simultaneously supported the panzer formations in their blitzkrieg. Polish cities were bombed to break the will of the people and their ground forces especially air force was virtually destroyed on ground. This was the first modern example of air force playing a decisive role in a victory.

Norway; 9th April 1940

Air Force as a whole is a strategic power; the value is in the capture and destruction of the target which fulfils the war aim. Norway was important for Germany because of its iron ore and the ability to attack the British fleet from air. But in the presence of Royal Fleet its capture by Germany was never given a serious thought by the Britain. It was the strategic use of the Luftwaffe that turned the table. The air assault and use of airborne troops by the Luftwaffe caught allies by surpris and Norway was captured by Germany despite the presence of

76 Scot Robertson, *Development of RAF Strategic Bombing Doctrine 1919-1939*,{London,Praeger,1992] p xx11.

strong British naval fleet and further blockading of Norway was hindered by the Luftwaffe when it attacked the them, which was forced to retrieve back to safe waters under the protective umbrella of coastal fighters.

Eben Emael; 10/11th May 1940

The Belgium defence revolved around strategic complex of forts of which Eben Email was the most formidable, and from German point of view their over all strategy of offense centred around its capitulation, one way was to destroy it from aerial bombing, but it was lengthy and not result oriented approach, thus Luftwaffe troops stormed it with gliders and captured it intact. This gives a new dimension to strategic airpower, where it is not only meant to destroy a target but to capture. The strategic result was overwhelming. It was a classic display of air power not revolving around expensive and heavy bombers but on wooden gliders.

Rotterdam; 14th May 1940

The Dutch resistance was broken by employing Douhet's doctrine of destroying and burning a whole city. Although there is enough[77] historical evidence points that, it was not the intention and Germany was abiding the commitment of 18 September 1939 given to the American President for not attacking the civilian population, yet the end effect was apparent, a state can be brought to its knees by sheer volume of destruction to its principal city. Germany had only 4500 parachute troops and only 4000 were used in attack on Holland, they were backed up by a light infantry division consisting of 12000 men carried in transport aircraft, in the words of General Student 'we dared

---

77 Liddel Hart History of Second World War{London,Cassell,1977} p72.

not fail, for if we did the whole invasion would have failed'.[78]

This was a new face of air power, away from the mere bombing and destruction of targets but capture of them to facilitate the over all strategic aims, thus it justify the inclusion of airborne troops as part of airpower.

Existing published material has overlooked the trials of AOP in the France with a statement 'experiment had been launched too late'.[79] Study of primary sources reveals following.

D Flight On 22nd January 1940 Major General Massy now Deputy Chief of Imperial General Staff{DCIGS} wrote letters to the Air Ministry and British Expeditionary Force {BEF},[80] both letters were contradiction to each other, one to Air Ministry stressed that the issue of Flying Observation Post {FOP} has taken a lot of time and it is understood that no suitable aircraft is yet found so another trials should be carried with a request that Taylorcraft aircraft should be modified for this by 29th January 1940. Other written to the BEF 'the machine is excellent…now only it has to be tested under battlefield environments whether it can evade the enemy aircraft'.[81] Thus at this stage DCIGS hoped that aircraft will be selected on one hand and on the other hand he is informing the BEF that it has been already selected. This selected dissemination of information suiting the over all concept of Air Observation Post was a double edge sword as it created a general confusion among all the concerned staff officers.

On 26th January 1940 RAF agreed to holding the trials in France, amidst the expected chaos of the war the AOP required three section of No. 87 Squadron RAF for protection.

---

78 Ibid ,p 73.

79 Parham and Belfield, *Unarmed into Battle*, p.16.

80 WO 197/70 B.M. M.O.7./15. dated 20th January 1940,also letters dated 25th April 1940 and MAF document No.14 Folder No 5.

81 Ibid dated 26th January 1940,AIR 35/111 S 1411/air 4 dated 1st February 1940.

On 31st January 1940 Wing Commander Blackwood wrote to British air mission in France highlighting the forthcoming trials 'a short period of trials may be undertaken {at Old Sarum}before flying to France'[82] RAF also agreed that 139 Squadron will provide the aerial cover by using Bleinhem and also to simulate enemy fighter attack.

On 2nd February Captain Bazeley visited France to see how French Artillery is experimenting with aerial control on 8th February he wrote the report in which he highlighted that French are carrying out the similar tests but the major difference with them is that they are using Autogiro and moreover in their concept the aircraft is not flown by the gunner officer rather he is the observer.[83] Basing on this report Massy issued detail instructions for the trial of FOP in the France which was already proposed by the No 22 Group. The aim was to test the aircraft, to contemplate the organisation of the FOP and to carry out live shooting. In the first phase the flight was to be stationed with the BEF with an entourage of an artillery and anti aircraft battery to proceed to Maillay for trials with the French and on 9th May this flight was to report at the 51st British Infantry Division, this division was instructed to keep this flight as far as possible in the French zone of 3rd Army because the Luftwaffe is bound to attack it,this division was moved into the French 3 Army area to relieve the 7th French Division. In BEF there was no idea about this AOP as one liaison officer wrote to the other 'Bazeley flying circus is coming ,I am not sure what it is'.[84] Massy did not inform RAF and neither he himself proposed that British Army should also put in to practice the French concept of AOP where observer is not the flyer.

Air Ministry handed the task to No 22 Group which tasked No 77 Wing who formed a Flight No1488,to carry out the experiments[85].

82 Letter dated 26th January 1940 from Director of Military Co-operation,AIR 35/113 also see AIR 35/111.

83 MAF. D Flight papers ,Folder No 5,Document No.15

84 AIR 35/113 letter of Captain Koch De Gooreyand,Liasion Officer,dated 2nd March 1940

85 AIR 35/111; letters 1411/air 4 dated 26th January 1940;and letter dated 6th May 1940

But the first and foremost issue was the aircraft. The existing aircraft which were under study included American Stinson, Taylorcraft and Cagnet. RAF earnestly wanted to develop this FOP procedure as it would have spared their crew, they were under the impression that if this light aircraft is inducted than the army co-operation aircraft would be freed, they recommended Tiger Moth aircraft. On 22nd March 1940 for security reasons the nomenclature of the flight was changed to D Flight, the contention of the Canadian Air OP history is wrong which suggests that D Flight name was given because of Taylorcrfat aircraft. War Office also decided that in future instead of Flying Observation Post the nomenclature would be Air Observation Post{AOP}. There is no record of any discussion on this issue and neither it has ever been debated that why this change of name should take place.

In a conference held at Colonel's Chilton office it was agreed that raison d'etre of the experiment which the official paper has termed as 'travelling Circus' is to produce a Flying O.P. in the Maginot Line and to operate it in conjunction with a battery or batteries of RHA over a period of three weeks approximately, it was expected and appreciated that Luftwaffe will attack the outfit but it was also anticipated that 'ground defence against such attack will inflict such severe losses on the enemy fighters that these attacks will not be repeated'[86] how naïve were the British.

The tentative dates discussed and approved were March 10th -23rd March 1940 to practice at Arras, between March 24th to 15th April at Mailiy and at SAAR from 14th April to 7th May.

On 21st February the C.in C. ,B.E.F. wrote to Commandant en Chef sur Le Front NORD-EST and informed him about the forthcoming trials of the Flying OP and stressed that British would be carrying out the

86 Secret Document .....

trials independently and hoped that French.

On 22nd March 1940 Chief of general Staff British Expeditionary Force wrote that nomenclature of the flight has been changed to D Flight for security reasons.

On 13th April Lieutenant Colonel Peppe of 1 Regiment Royal Horse Artillery decided that rendezvous at Mailly would be hotel Lion D' OR at Rheims at 1215 hours on 15thApril,atmaillyGeneralDeVaucelles Director des dours de tird;Artillerie and Colonel Prevot Commandant d'instructiond'ObservationAeriennedʼArtillerie on behalf of French army coordinated with D Flight.

The strength of D Flight was 5 Officers and 31 Airmen supported by another 13 officers and 429 other ranks. It had three Taylorcraft Plus and one Cagnet with a provision that one Stinson 105 may be added later.

Officers stayed at Hotel Carillon at Arras and 'other ranks accommodated in hired premises at 34 Rue d' Amiens where cooking, dinning and sleeping space exists, rations and fuel will be drawn daily from ration store...bedboards and tables will be supplied under arrangements made by officer commanding headquarters unit.[87] This is extraordinary in nature because seldom troops marched to battlefield while staying in hotels and hired accommodation.

D Flight arrived in France with instructions to remain in Maillay from 1st May-11 May 1940 and then moved to 51st British Division area in SAAR. D Flight did arrived in France and on 14th April 1940 was stationed at Arras aerodrome with five officers and 31 air men, the flight preferred to stay at Hotel Carillon it moved to Maillay with an entourage of over 400 all ranks,[88] 73 vehicles with one battery of Royal

87 WO 1078.HQAC/623/32/ORG dated 14th April 1940
88 D flight had 16 officers and 396 other ranks,three Taylorcraft,one Cagnet and one Stinson aircraft.

Horse Artillery and one battery of Light Anti Aircraft less one troop which instantly demanded camouflage nets and certain administrative set up, the anti aircraft battery was later replaced by another battery comprising of eight Bofors guns ex 51 Light Anti Aircraft Regiment which also put up demands of having concrete pill boxes for safety against enemy air attacks. The Bofors were instructed to have careful digging and concealment of the positions otherwise being so far up they will be engaged by enemy artillery and neutralized, an efficient warning system to inform the command post of approaching enemy aircraft. They were all under command Lieutenant Colonel Hilton Peppe of Royal Horse Artillery with Major {squadron leader} Bazeley was given a carte blanches to communicate direct to War Office. The general impression that arises out of this initial spectrum is of an army which lacks will to fight and is totally unprepared for a war and where the officers at all level have failed to comprehend the gravity of the time for them the personal and regimental ego was far important than the overall art of war. War Office also wrote to BEF on 6th March 1940 explaining them that the Flight has to have artillery pilots and record exists of pilots from Royal Artillery having A & B flying licenses. French has also agreed on the trials of the D Flight and gave them the frequency bandwidth of 4225,4500 and 4700 Kcs.

D Flight had T.R.9B wireless set which was not crystal controlled and could work in the frequency band of No.11 Army set it also had T.R.9.D which was crystal operated, on 22nd March 1940 the signal officer in chief had warned that army set No 11 had to be provided in sufficient numbers to battery posts and observers for anti aircraft protection and also for communication with flying observation posts. Thus it is surprising that D Flight did not conduct any operational sortie[89] nor any artillery shoot neither they were utilised by the BEF and

89 AIR 2/7188. D Flight War diary2r/gD1/1,Form 540,see also D Flight report of 13th May 1940 ,MAF Document No 19.,box no 5. also see WO 216/77 letter dated 5th may 1940 also see the *History of the 15th Scottish*

nor they themselves volunteer to do so. By 7th May 1940 rumours of German onslaught were at their heights and finally on 10th May 1940 the Blitzkrieg was on the roll and aircraft were moved by Baizeley from the aerodrome to close by open areas and like rest of British and French armed forces he too was mentally paralysed, otherwise there is no other plausible explanation for his not flying to the BEF area especially the 51st Highland Division and offering his services, his other mentor Lieutenant Colonel Parham was commanding a medium regiment there and he should have been happy to receive him and utilised his services. Bazeley send the RAF segment of the D Flight to make their way at their own and himself and other pilots flew to Dieppe and from where he send an officer with an aircraft to fly back to London and informed the War Office and also to get the signal issued for their return, Captain Davenport flew to London on 19th May and very next day the rest of the two aircraft also flew back without any escort across the channel to Old Sarum. Miraculously the RAF party also managed to get back home without any mishap leaving the attached artillery and ack ack to fight their way back. This episode bring into light something which was not contemplated by the gunners ,they had only concentrated on the part where the battery commander is able to fly his aircraft for observation parked next to his tent but in case of a withdrawal as happened now the military ethics and moral duty of a battery commander lies with his troops and his or any other officer of the battery flying out and leaving the troops certainly lower their morale. The future relationship of RAF lower ranks and AOP remained very strained and in certain cases these RAF ranks simply left them during exercises at home,[90] Air OP historians and officers have been very critical of this conduct of RAF yet the roots were laid here in France because of Captain Bazeley's self eccentric conduct. But it was not unusual as at least one RAF squadron also simply left the

---

Division 1939-1945 by Lieutenant General H.G.Martin {William Blackwood & Sons,Edinburgh 1948}.p-32.Also see the A Short History of No.652 AOP Squadron,MAF.

90 Lyell ,Memoirs of Air OP,p.32 .

airfield without even waiting for its fighters to land back.[91]

The sudden collapse of French military is one of the most discussed enigma of modern military, how come a nation which is in state of war since September 1939 and yet fails to read German war plans, no mass air raids were conducted neither any city was burnt like Rotterdam yet the whole military collapse. It was the speed of the German offensive led by the Luftwaffe the difference between tactical or strategic airpower is blurred. Luftwaffe achieved collapse of enemy without concentrating at its industrial base. It was lack of will to fight by the French that was the main cause of their downfall. RAF's and infact the whole british strategy collapsed, the knockout blow failed and so did Roayal Navy.

Over 3,00,000 British Expeditionary Force was evacuated from the Dunkirk (26th May-3rd June 1940) by the British Navy within nine days, Churchill called it 'Miracle of Dunkirk', however it was the failure of Luftwaffe to destroy the forces and to wretch the air superiority from RAF which resulted in this miracle, although the fog at Dunkirk played an important role in this, yet the fact remains, that out of two air forces ,RAF and Luftwaffe, the RAF was able to resist and fight in the air ,thereby accomplishing a strategic task. Again if one see the things from other angle, and assume that if at Dunkirk Luftwaffe is able to hold off the British Fleet and RAF, then probably the war would have ended there, thus it was failure of one air force and success of the other but still it was the airpower that was the deciding factor.

The experience of the France had a deep impact on both services as both blamed each other for the fiasco The consensus among the army was that RAF has let them down at least when compared to the Luftwaffe's support to their army which in any case was not off the mark.[92] Bazeley's

91 MD/2242:*Brown With Two Shades of Blue, Reminiscences of an air observation gunner service with RAF and Fleet Air Arm*,p-3,4,9,14,15

92 AIR 2/7188;WO letter 0144/7551[M.O.7] 11th July 1940 regarding the provision of tactical and

report shows that AOP was unable to carry out the observation beyond seven thousand yards, communication was the major irritant. All the pre war concepts which Bazeley had himself written about the AOP now proved in vain; biggest reality was that it {AOP}cannot operate under hostile air environments, it requires sufficient air cover and ground protection apart frcm need of an anti aircraft battery, all relics of WW1.

Fate of the D Flight was decided initially by the RAF to let it live and soon War Office took charge of the affairs, thus the contention that RAF was against the AOP is not justified. However it was again General Massy who was now Director Military Training to keep the concept of AOP alive he ensured that AOP be given another six months during which it was placed under the Home Forces and a questionnaire was prepared to be answered by the army formations and artillery school on the validity of the concept. The fear of invasion was another key factor for the survival of D Flight. There is no War Diary of D Flight it self in France and its war diary and digest of service starts from 1st June 1940 which mentions the arrival of ground party at Oldham which was a RAF station under command of P/O Forth and aircraft remaining at Old Sarum[93].

D Flight had a busy schedule Bazeley and Hilton put up their best to make this concept survive. The questionnaire had ten questions and dealt with the theme to find how far AOP can observe the target, can it differentiate between own and enemy infantry and what are its chances of survival in hostile air environments. It was also decided even before the trials were completed that another batch of artillery officers should be trained as AOP ,thus over fifty officers were earmarked to undergo course of six weeks at elementary flying training school and then further six weeks training at Larkhill with D Flight, this clearly indicates that War Office had already made up its mind that AOP is useful without

strategic aircraft for army in the field.Army air requirement ,note by Chief of the Air Staff 1st November 1941

93 Form 540,D-Flight IIr/gd1/1,MAF.

waiting for the report. Otherwise one of the question that whether it can survive under hostile air environments was answered in the France.

The selection of an aircraft for AOP still remained a thorny issue with RAF finally reaching to the conclusion that there is no aircraft which full fills the AOP requirement,[94] Major Bazeley had already rejected the Taylorcraft during trials in France, again on his recommendation the American Stinson aircraft was recommended. In America during 1930 the first use of aircraft for artillery observation were carried out by the second lieutenant Joseph Mcord Watson with 36th Diviosnal Artillery[95] and in 1939 American Army conducted a design competition for small liaison aircraft in which Stinson 0-49 and Ryan 'Dragon Fly' stood the test. On the advice of Major Bazeley War Office placed an order of 100 Stinson aircraft which were to arrived in June 1941. Meanwhile RAF did offer even Hurricane apart from Blenheim to be used as the AOP aircraft but they were rejected. This was the cardinal sin to reject a fighters to be flown by the gunners, fighter would have been ideal; War Office instructions for the use of air craft given in WW1 still stood true. it was the direct method which is best suited for the moving battle.

June 1940, on 10th June 1940 Italy entered into the war as a friend of Germany ,it resulted in the closure of the Mediterranean, Italy had over 2,50,000 troops in Libya and another 3,50,000 in East Africa. Air Chief Marshall Longmore AOC-in-C Middle East, his command spread over all RAF units in Egypt, Sudan, Trans Jordan, East Africa, Aden and Somaliland, Iraq, Cyprus, Turkey, Balkans, Red Sea, Mediterranean Sea,Persian Gulf an area of four and half million square miles,he had 29 squadrons under his command.[96]

---

94 AIR and WO

95 MR/192 Report of the General Board United States Forces European Theatre report on study of organic field artillery air observation {study No 66}pp.1-39.

96 Right of Line,p-304.

Battle of Britain; July-November 1940

The battle around which the fate of British Empire rested around in the hands of few thousand men, and this is also the true definition of a air power. Germany after its victories on mainland Europe gets frustrated at the defiance of Britain, for its consistent political resistance, thus there was no other way except to either invade the island or to destroy it industrially and morally to an extent that it cease to exist as a formidable foe, and for both eventualities it is the air, from British point of view their whole existence revolves around this, can they face the wrath like Rotterdam.

The first phase of battle started in July 1940 with Luftwaffe attacking in a haphazard manner it lasted till 3rd August, the targets were naval harbours and convoys. Luftwaffe lost 364 bombers and fighters, RAF lost 203 fighters.

The real onslaught came in first week of August 1940 in which the RAF airfield were targeted in order to gain air supremacy, the high point reached on 15th August 1940 when Luftwaffe launched 1786 sorties in which over 500 bombers were employed, airfields in north and south of England were targeted, Luftwaffe lost seventy five aircraft as compared to thirty four RAF fighters. It was this day that inspired Churchill to say 'Never in the field of human conflict was so much owed by so many to so few'.[97] Statistically Luftwaffe lost 167 aircraft between two weeks from 10th August 1940 onwards as compared to 174 fighters of RAF.

The Third phase started on 24th August 1940 and this aimed at hitting RAF facilities around London, on the same night some ten bombers of Luftwaffe dropped [98]unintentional pay load over central London. Very next day some eighty British bombers attacked Berlin followed by more,

---

97 Ibid p 107
98 Ibid p 109

leading Hitler, after threats that were ignored, to order reprisal raids on London. Between 7th September-13th November 1940 Luftwaffe flew over 12000 sorties over 13000 civilians were killed and another 20,000 were injured in London alone,[99] yet morale held up surprisingly well. Attack on Coventry on 14th November killed 380 civilians and injured another 800 It is at this point that the strategic air power assumes what it was designed from the beginning, to terrorise the population and to break the fighting will of the people, it all started accidently but from this point onward there was no stopping and became a standard pattern in the rest of the war. Thus bombers became the symbol of aerial menace over cities and attacks over unarmed population became an integral part of strategic air power .

Army Cooperation Command. In November 1940 was set up under Air Marshall Barrett and Air OP was placed under him{as part of training group under No 70 Wing}. This was a highly desirable and required step as for the first time RAF accepted the need of army requirements, from AOP concept it was a turning point, they came under a full fledged command of RAF which eases its administrative and training requirements.

British Army had decided to raise 15 armoured divisions and as such they put up a requirement of 60 close support squadrons in which Lysanders were to be replaced by Tomahawks and Blenheims 'I am further convinced that unless the complete control of these squadrons is vested in the army, it will be impossible to ensure that they receive the necessary training'.[100] Close Support was first conducted on an inter service basis in August 1940 which resulted in Vahn-Woodal report, it was accepted by the War Office & Air Ministry. The British were now going back to the basics of art of war, their very cardinal concept of

99 Roy Nisbet An illustrated History of RAF,{London,Salamander,2002} p 144
100 AIR 39/16,ACC /44/9/AIR air support for army ,GHQ memorandum and WO memorandum.

winning the war through bombers had failed miserably and thus the reversion to close support took its reincarnation for which RAF was not willing by heart. On 4th December 1940 it was agreed that all medium bombers will under go close support training and on 10th December 1940 it was also agreed upon to have 500 such training sorties however by end March 1941 only 45 such training sorties took place.

The report[101] at the end of the year from home forces, artillery school and RAF No 73 Wing were all favourable. RAF did point out that only '30% of observation is possible from ground observation' and the balance cannot be overcome through the army co-operation fast moving aircraft which were being used for the artillery reconnaissance, thus AOP seems to be a viable solution for sharing the load but it will not replace the army cooperation aircraft .Army formations on the other hand were also favourable for the AOP but their analysis were similar to the trial reports of early days The most glaring part of the trials were that AOP can only observe up till 3000-4000 yards, it requires fighter protection, it requires ground & ack ack cover, it needs very close coordination with the supporting artillery battery, 'AOP is useful but it cannot meet the long range artillery requirement'.[102]

Air Marshall Arthur Barrett, is generally not regarded and remembered in good words by Air OP officers and historians[103] he is ridiculed for his remarks about the AOP where he termed the concept as unviable, but the AOP has been unfair to him and his quotes and remarks have been taken and quoted out of context. In fact none has helped AOP more than him ,he was the first one to analyse the whole idea from a professional point of view. Once the southern command and artillery school had given favourable remarks for the AOP it was Barret who was asked to forward his, and in his two reports he elaborated the idea

101 Ibid
102 Ibid
103 Memoirs of an Air Observation Officer pp and Unarmed into Battle pp

by stating that whole concept is revolving around ability of an aircraft to take off and land close to the artillery battery but Arthur pointed out that this is possible only in the plains of Salisbury, but what will happen if this AOP is deployed outside homeland where there is no such area, further he went to highlight the tactical dilemma in case if the battery is attacked by infiltration as aircraft which can not fly at night would be a liability, moreover it cannot fly in bad weather. Barrett's judgement was that AOP is good enough for homeland but not for abroad. In his first report he did not committed himself to any definite answer he was asked again to be precise and that is when he wrote. '*The case against the AOP is final, it cannot stayed alive…Artillery should explore the other means including Kite observers*' and it is these remarks that are always exploited by the AOP. Yet the hard fact is that artillery as a whole never ever explore any other means nor even gave a try; further more in the time period starting from 1938 the other side of the issue that RAF pilots cannot be trained as gunner was never contemplated. It is worth pointing that Barrett himself was commissioned in artillery in 1910 and served as artillery observer in RFC during WW1 thus he had more vision than any of the military officer on the issue. Director military training Lieutenant General Massy countered all the negative remarks by stating that 'we are preparing for an European War and it should be seen in this contest'[104] he further highlighted that there are over 187 artillery field regiments and 27 medium regiments at home and over 56 abroad ideally there should be an aircraft with each one of them. Massy at least came up with the exact figure and the role of AOP although by stressing that it is meant only for the European Theatre he admitted the inherent flaw of the idea. On 18th April 1941 a conference at Old Sarum was held by the AC Command to iron out the procedures, the aim was to increase the efficiency of the squadrons ' there is a tendency to be absent without official leave{AWOL} during parade hours offenders are not brought

104 WO 32/10413

to the tasks…neither discovered…loose talk about Lysander is bad for morale'.[105]

Unfortunately Barrett's remarks came up at a time when army was pressing hard for more and more RAF support for ground operations and RAF was resisting as hard as it can.[106] CIGS John Dill on 30th May 1941 pressed for almost 109 squadrons[107] to be earmarked for its requirements, among this list the last item were requirement for AOP Squadrons.

'To enable the artillery to carry out air shoots at very short notice quickly and efficiently an organisation for the observation and direction of fire, consisting of Air Observation Posts in the form of squadrons of light aeroplanes with royal artillery pilots is required. The immediate requirement is for three squadrons, the aircraft to be supplied in the first instance from any light aeroplane resources which are available. These squadrons will later be re-equipped with the one hundred Stinson O-74 aircraft which are on order from America and whose delivery should be hastened. The ultimate requirement is for twelve squadrons each of twelve aircraft'. CIGS had asked for a total of 3888 aircraft for the support of the army which was almost 300 aircraft more than the existing first line strength of RAF.

RAF on the other hand strongly contended this and cited that Luftwaffe support for 154 German divisions was only 3000 aircraft 'Germany does snot possess a large specialised air component…the success of German Army has been won by selecting a battle field which places the opposing air force at a strategical and tactical disadvantage…

105 AIR 39/32 conference Air Co-operation Command held on 18th April 1941 at Old Sarum.

106 AIR 39/16 ACC/44/9/AIR;Air support for Army.GHQ and War Office memorandums

107 AIR 39/16 ACC/44/9/AIR;Air support for Army. GHQ and War Office memorandums, the distribution was to have 3 squadron at home for each corps and 3 per armoured division, at overseas 5 squadron per corps and 5 squadron for armoured corps,thus 54 squadrons at home and 55 squadrons overseas.half of these squadrons to be equipped with fighter reconnaissance and other half with bomber reconnaissance.

army not the air force is Germany's primary weapons'.[108] On the issue of Air OP, Air Chief later wrote 'provision of AOP squadrons should remain on low priority as they take equal resources as any regular army co-operation squadron demands'.[109] The revised requirements of the army were put up again and these were discussed and RAF finally gave way on the AOP issue.[110] The other notable event of that time was the raising of the regiment of RAF in which guns were also included and RAF requested for army artillery instructors there is no reason to doubt that the raising of AOP squadron was as a barter.

Thus in this haste the first AOP squadron was raised on 1st August 1940 at Old Sarum, it had no aircraft and neither any organisation. War Office informed the southern command on 10th August 1941that first squadron is now raised and its nomenclature is No 1 Air Observation Squadron Royal Air Force {Army Personnel} however it was RAF which allotted the block serial of 600 to AOP squadrons and thus this first squadron was christened as No 651 AOP Squadron RAF{Army Personnel}. Squadron practically evolved out of the D Flight, where as D Flight remained alive and was later converted into 1484 Flight,for training purposes.

On 29th September 1941 Exercise Bumper was conducted in which Lieutenant General Massy was the corps commander, 54th Division and 86th Field Regiment also took part in it and so did the first AOP Squadron but the relationships were so soured that RAF men refused to work in military way by highlighting 'own time and government time'[111] later on 7th October 1941 Exercise Percy was held in which one flight of the squadron took part; Captain Jim Neaderthal turned upside down

108 Chief of Air Staff memorandum dated 1st November 1941. AIR 39/16 ACC/44/9/.

109 AIR 39/16 ACC/44/9/AIR;Air support for Army.GHQ and War Office memorandums,paras ,air chief's reply dated 17th june 1941.also see Army air requirement.

110 AIR 39/16; ACC/44/9/AIR and acc/s.142/air.;Air support for Army.GHQ and War Office memorandums dated 12th may ,paras 11 and 12.

111 Memoir of an Air OP pilot.pp,29-34.

and Captain 'Monkey' Morgan got burnt due to hitting the high tension wires. The communication system of the AOP was very local and only to the army ground sets which soon lost contact in cross country navigations. AOP pilots felt more at home with RAF than with army 'RAF understood the pilots far better than did the army'[112] where as in the army the leave was given for one week after 12 weeks the RAF was giving the same duration of leave after six weeks, moreover the RAF air crew ration was much more than the army ration.

The evaluation of AOP on the home land during the manoeuvres were praised by the formation commanders and it was rightly so because on home ground AOP was able to operate from the damp strips{as English cricket team does} and provide limited artillery support its major impact was in the transportation of senior officers which before it were relying heavily upon on the RAF and seldom got that thus they were able to save time now with the AOP and it was all against the rules. Home Forces Commander had to issue strict orders that AOP's use as a taxi should be avoided.[113] The favourable reports which AOP got for its continue living are mostly due to this performance. The command channel of AOP was also another source of irritation as they were under the control of the Commander Royal Artillery and it was his discretion to allot the aircraft and sorties which RAF objected as they usually were not in line with the flying regulations.

112 Ibid.p-14.
113 Ibid

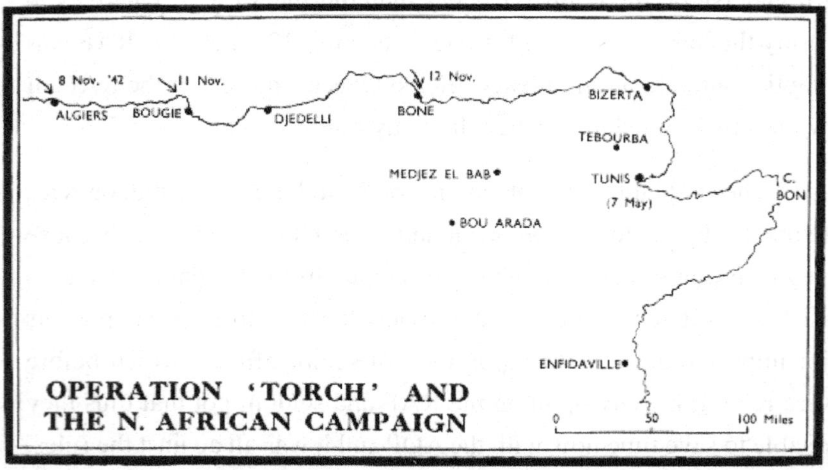

Maps Courtesy "UNARMED INTO BATTLE".

Chapter Three 1941-1943

Air OP Campaigns

Africa & Italy

In August 1940 Air Chief Marshall Sir Fredrick Bowhill advocated the formation of a ground defence corps,from 1940 onwards RAF had a new trade 'gunner'. RAF regiment was raised on 6th January 1942[114] as a corps within the RAF there was inevitable opposition to this by War Office, it later expanded to over 50,000 troops in 240 anti aircraft,infantry and armoured squadrons having 75 mm Howitzers,25 pounder guns, three inch mortars to name few. One of the key reason for formation of this regiment was that after its formation over 90,000 army troops were released from the protection duties of RAF airfields, it was accepted by War Office on 1st December 1941.[115] It is surprising that Air Observation Post Squadrons were not made an integral part of Army Air Corps which was also raised in 1942.

In April 1941 Germany attacked Yugoslavia followed by Greece and within days three allied divisions and an armoured brigade were routed, yet another expeditionary force had to be evacuated and whole equipment was either destroyed or captured by Germans. Belgrade was bombed on 17th April 1941 which forced the government to surrender. 'In this fighting German air superiority had played as significant role as it had in the Blitzkrieg in Norway and Western Europe;[116] Parachutist again played an important role in the seizure of Corinth Canal in Greece.

114 Oliver Kingsley,*RAF regiment at War 1942-1946*{Leo Cooper,Yorkshire,2002},p-6.
115 Oliver Kingsley,*RAF regiment at War 1942-1946*{Leo Cooper,Yorkshire,2002{,p-14
116 Gerhard Weinberg,A World At Arms,{United Kingdom,Cambridge University Press,1999} p 226

Crete; 20th May 1941

An island whose occupation by any belligerent gave it a strategic advantage for the war in Africa, it was defended by Royal Navy, yet on 20th May 1941, first wave of 3000 German parachutists started landing on the island held by 28,600 allied troops along with two Greek divisions and by 26th May evacuation of allied troops started and by 1st June 1941 over 16000 British troops were evacuated with equal number captured by Germany as prisoner of war, it was a classic display of what airpower can achieved, it was a combination of fighters gaining air superiority, bombers bombing and finally the parachute troops capturing the air fields and follow up troops landing, there by totally making navy redundant. It was high mark of air power. 'Capture of Crete by an invasion delivered purely by air was one of the most astonishing and audacious feats of war'.[117]

Barbarossa; 22nd June 1941

Germany attacked Russia on 22nd June 1941 and by December same year reached the gates of Moscow, Leningrad and Stalingrad, before being halted, once again the Luftwaffe played an important role in knocking out Red Air Force on ground, there by ensuring safe passages for panzers. It was however the decisive role that was played by Luftwaffe in the Russian counter attack in the winter that trapped over 90,000 German troops of General Walter Graf at Kholm and at Demyansk, Luftwaffe was able to evacuate over 36000 wounded and also supplied over 30,000 tons of vital supplies, and 31000 troops were replaced, Luftwaffe lost 265 aircraft and Red Army lost 408 aircraft,[118] but

---

117 Liddel Hart,History of WW2 {London,Cassel,1977] p 141

118 Harold Faber,Luftwaffe an Analysis by Former Luftwaffe Generals{London,Sigwick,1979} p 221

it was the first instant that an air force was able to prevent a catastrophic chaos caused by encirclement of own troops by the enemy. One has to compare the role of Luftwaffe and RAF in similar circumstances; RAF failed to do so in France May 1940.

On the other hand lack of a four engine long range bomber became a major strategic obstacle as Russian industrial power house was outside the reach of Luftwaffe bombers, thus Russia was able to feed its army with vital supplies that outpaced the destruction of German air and land firepower. Moscow was bombed but without any strategic impact. The fate of 6th Army at Stalingrad, finally hinged around the ability of the airpower to feed it and evacuate it but Luftwaffe was unable to do so, thereby lack of long range aircraft became one of the leading cause of German defeat and its final destruction in the hand of red army and air force.

Myth of Aryan was shattered and counter moves from allies started which began with a push in Africa and Italy.

Operation Torch. Britain had a considerable presence in Egypt for the protection of Suez and so did Italy which was occupying Cyrencia, Somaliland, Ethiopia and East Africa after Italy's entry into the war the flames of death spread to the Africa as well, British soon pushed Italian back and in January 1942 Erwin Rommel made his entry and turned the tables on British with some remarkable display of leadership qualities. The situation remained fluid however by mid 1942 Allies planned an invasion on North Western Africa {Algeria, Morocco and Tunis}as a check to Rommel's offensive drive.

North African invasion's preparation started in early part of the year 1942 ,on 13th August 1942 the newly formed H.Q. 1st Army moved from Scotland to London and started the preliminary paper work

Brigadier Parham CCRA 1st Army; one of the early proponents of AOP was also involved in the planning phase he recommended the induction of AOP in the assault troops. He admits 'no one in the planning had any clue about the AOP... AOP to be used till it finishes it self'.[119] Bazeley was also attached with the Parham for planning. The raising matters of AOP became complicated with every passing day, artillery wanted to stamp its own identity and demanded that the vehicles of AOP be marked with army tactical numbers resultantly the AOP Squadron got both RAF and army numbers, army gave Mob serial No 34506 and RAF number was A9189,this rather inconspicuous act had severe retarding effect later as services rules restricted the drivers and transport to carry their own services load, yet another eccentricity of British military system many a times army drivers refused to carry the load for the squadron because the marking was in RAF serials meant for the air craft and similar was the reaction of RAF drivers. Tactical aspect of AOP was never considered in the army manoeuvres, the organisation was based upon the RAF pattern, in fact it was Air Marshall Barrett who gave the first out line organisation but it was based upon the preamble that AOP would be operating from the static bases of the RAF. The relation ship of other ranks of RAF and AOP army officers were highly demoralising with RAF troops in certain cases leaving the pilots without even informing them to spend the night in the comfort of the air base. No. 651 Squadron was earmarked for the invasion but surprisingly no close co-ordination took place between the pilots and the affiliated regiments of artillery one of the fundamental principle of AOP. It is no surprise that AOP was the last unit to get the orders for mobilisation on 12th September 1942 with orders countermanded no less than three times. Major Bazeley has blamed RAF for this delay but the root cause was the twin channel of commands 'conflicting orders from both services were sent to the squadron'.[120]

119 Ibid
120 WO Major Bazeley's report administrative Problems,2/Air/5.April 1943.p.1 para 2

Before the onset of North African campaign the AOP had drifted too much away from its original concept which the artillery and army were still presenting to the RAF yet in reality it was not there. Instead of mature artillery officers of battery commander calibre these were young officers who had little knowledge of military procedures, as they were still being taught in the AOP courses the drills and tactics apart from the layout of the division.[121] The other pillar of AOP was the battery commander who should be able to take off from the ground next to his battery and carry out the fire correction but now the aircraft were not attached to the batteries rather to the regiment and pilot had no intimate knowledge of the guns layout or the actual tactical disposition. The aircraft selection remained a thorny issue, the original Taylorcraft around which the concept was evolved had vanished in the favour of the American Stinson aircraft and over 100 were ordered but the very first batch had the mishap of being transported under the cheese crates and arrived damaged,[122] the further supplies were in doldrums as American Army itself selected the same aircraft for its liaison duties.

On 1st May 1942 second Air OP squadron No. 652 was raised at Old Sarum with an establishment of eighteen officers from Royal Artillery and two officers from RAF with 130 other ranks of both services commanded by R.R.Cobley equipped initially with six Tiger Moth air craft re-equipped with Auster Mk1 and before the end of the year with Mk3. Squadron moved to Bottisham soon after forming and carried out its initial training under BRA Eastern Command, in August the squadron moved to Westley, Bury St. Edmonds and by December the squadron was in Dumfries where a special flight 'S' was raised under Captain G.A.Hill to support Force 125 for an operation which was later cancelled. In December 1942 Derek Heathcote-Amery from War Office

121 MD/4024 Creswell,Lieutenant Colonel John Hector.*Collected pilots notebooks,course note book, manuscript diary of 7 Section,B Flight AOP Squadron RAF,pilots flying log and two published officers Association Reviews.*

122 Ibid

who later became the chancellor of exchequer spent a week's leave visiting the Air OP at Old Sarum, he was given the joy rides 'I got the job of teaching him and he was delightful pupil in every way…after Derek's visit to us he thoroughly understood all of our needs and we had no further problem with the War Office'.[123] In December 1942 Lieutenant Llewellyn flew into high tension electric wires while flying low ,he was killed, he became the first officer to die in training in Air OP. The life of army pilots was not boring or just routine.

The role of staff officers is very important in any war at least in fostering the relationship between the services. Artillery advisors unfortunately fell below this mark as they added fuel to the fire by disseminating information to the commanders which had no truth as is obvious from the following letter. On 18th February 1942 General Officer Commanding Home Forces General Bernard wrote a demi official letter to the C.I.G.S. General Alan Brooke, the home forces commander's three page letter was harshly worded he wrote,

'M.G.R.A tells me…that certain officers in Air Ministry are the real hurdle…There is a calculated obstruction to getting anything moving…war office is not firm enough…I ask for further 13 A.O.P. squadrons'.[124] To close the arguments he stated the authority be given to take over the AOP by the artillery.

It took almost a fortnight before CIGS replied in black and white on 16th March 1942 he replied and right from the onset made it clear that there is no deliberate efforts on anybody's part to derail the AOP 'I think you have been misinformed…The issues at stake are very much bigger than these of Air OP squadrons…these Air OP squadrons takes a very low priority.[125]

123 Memoirs of an air observation post officer,p-51. strictly speaking from rules his giving flying lessons to Amery was not only against the rules but considered dangerous but it was the army way of getting the things done.

124 WO 216/77 letter from C-n-C Home Forces H.F./6308/1/R.A.{F.A.}dated 18th February 1942

125 WO 216/77 ;CIGS reply

Thus it becomes clear that its not only the RAF or Air Ministry but also War Office that viewed the Air OP not as a priority one. RAF had highlighted that it takes equal resources to raise AOP or fighter squadron, these AOP squadrons were not relieving the RAF from the responsibility of Arty R. However due to the pressure of the home forces commander resultantly the paperwork started to replace the Stinson with Auster aircraft and by May 1942 Auster 3 aircraft started their debut with the AOP. Thus it was army which accepted an aircraft which it had previously rejected on the grounds that it does not full fill the requirements of an AOP and now over 700 were ordered. This induction of Auster was again challenged by the AOC. Army Co-Operation Command Air Marshall Barrett on the grounds that he was never asked in the induction and above he cannot allow his pilots to fly an aircraft which is unsafe.

The AOP Flight after sailing from Glasgow docked at Algiers on 12th November 1942 in the follow up echelon with 'eight crated Auster aircraft eleven pilots, sixty four other ranks and twenty three motor vehicles and one staff car'.[126] Three days later flight assembled the four Auster aircraft and then flew amidst air cover to report to 1st Army which in reality was nothing more than a corps. AOP by its design was not meant to be so heavily loaded in terms of transport.

The tactical situation on ground along with weather restricted the deployment of the Austers, they were attached with the 78 Infantry Division and sections were further attached with each of the supporting brigade with flight headquarters remaining with the corps headquarters. The performance of AOP in this short campaign which unexpectedly turned in favour of Germans by virtue of their quick reinforcements highlights the inherent weaknesses of the AOP. They conducted a mere 37 sorties out of which only ten were in support of the artillery, their deployment with brigade headquarters put them away from the affiliated

126 Parham, p.28 also see WO Major Bazeley's report administrative Problems,2/Air/5.April 1943.

regiments, the 13 Medium and 132 field regiments{in original concept after trials AOP was meant only for the field artillery and not for the medium regiment.} Their only sortie of any worthwhile cause was to try and locate 5th Northampton Battalion which was apparently lost in mountains and pilot Captain Oldman despite locating them was unable to communicate as the wireless communication was not matching.[127] Official history has covered the lack of sorties to the fact that the mountainous terrain gave equal advantage to the ground observer thus there was no need to take to the air.[128] The diary of 7 Section B Flight depicts a picture of utter frustration mainly due to the fact that Flight Headquarters was located at rear and spares had to be brought almost daily, he further highlights the limitation of night flying and operating from forward strips, one aircraft crashed due to landing in darkness, there were further incidents of German aircraft attacking the Auster which in certain case were not successful yet RAF as a precaution stopped flying as they were unable to provide the fighter escort for limited duration of the AOP which was getting airborne for mere twenty minutes and that too behind own lines.

By January 1943 the squadron was complete in entity with three flights supporting three divisions including 6 Armoured Division and Bazeley was recalled to the home. In the all important battles of Tunisia the AOP was missing as they took no part in the artillery bombardment, one reason was that most of the artillery bombardment took place in the dark and AOP lacked the means to fly in the dark. On the basis of the 651 Squadron performance the fate of second AOP depended and had it been a realistic appreciation of their performance the second squadron 654 might not have arrived yet artillery formation send a signal to the War Office complementing the role of the AOP 'Air Op already an unquestionable success despite adverse conditions …Austers…

127 Parham.p.35
128 Parham,p.37.

endangered pilots'.[129] Thus 654 Squadron also arrived in March 1943 and was later shifted to the 8th Army, in the end none of these squadrons played any role worth mentioning in the artillery bombardment. Log book of a pilot for the month of April 1943 reveals just one sortie for artillery and rest of first fifteen days flying concerns either ferrying of aircraft or seeking information sorties.[130] An example of a sortie is given by a flight report which undertook on 23/24th March 1943 between 2345-0045 hours under full moon conditions the object given in the report is to register harassing fire at track junction 504414, which appeared in aerial photographs to be a route being used by the Germans for supply and the other object mentioned is to give Captain Tallents experience of shooting at night.

'I {one more AOP pilot as passenger} took off had no difficulty in locating the target from 3000 feet so ordered the troop to fire I missed the opening section's salvo added 400 which I also missed but heard the burst below me, I flew further away and had no difficulty in seeing the next round and continued ranging until I had verified a 100 yards bracket on the target, I ordered a round of troop fire which fell in the target area, I then ordered Target Toukabeur one round gun fire which fall slightly short on addition of 300 yards range brought into the area so I dived towards Toukabeur and flew over it as low as possible to observe movement, I saw nothing so flew on towards reconnaissance ridge down the Toukabeur Road while Captain Tallents searched the gullies with high powered binoculars lent by light anti aircraft, no movement was seen so I returned to advance landing ground'[131] There was nothing wrong with the sortie as in war these things do happen but it highlight that not much of difference was created by having a gunner in the cockpit of a slow moving aircraft.

129 WO,message no 087611 dated 9th December 1942 from RA 1st Army
130 MD/4024 Creswell *Collected pilots notebooks.*
131 WO ,Flight Report Sortie No 6 dated 23 March 1943

On the other hand these squadrons were used for the transportation of the senior officer and even the flight commanders were unable to change the pilots which were attached with the General Anderson as personal pilots. '4th April 1943 section to be used as 8th Army commander private taxi, most depressing news'.[132] It is safe to assume that it was basing on their this utility that the signals for their ability were issued.

Bazeley was posted back to the artillery school and he later wrote a personal report for the commandant in which he highlighted everything from the issue of uniform to the pay and rewards discrepancy and yet had no words on the tactical doctrine of the AOP in field 'in two months after arrival no RAF officer had visited my headquarters or any of my flight headquarters.'[133] The burning question was whether a flight should operate as a unit or should the sections be split ,the experience of forward deployment of aircraft with the artillery regiments had a negative impact as these units in a fluid battle situation were bound to change their location and with the inability of AOP to fly at night they became more of a liability as it happened on 29th/30th November 1940 when strip came under attack at Tebourba. Bazeley strongly recommended that the maintenance flight which was integral to the flight headquarters and consisted entirely of RAF personals' and commanded by RAF officers should be separated from the squadron as it hampered the squadron's mobility, however this is another indication of the relationship between army and RAF even within a small unit.

The over-rated signals about the performance of AOP were a major reason that no serious evaluation was undertaken for its improvement. BRA 1st Army praised the performance of AOP and later his signal was sent to India as well 'The Air OP really came into its own under battle

---

132 MD/4024 Creswell;.entry of diary dated 4th April 1943.

133 WO 2/air/5 dated April 1943

conditions, many regimental and divisional shoots were done on 9 Corps front with excellent results and in spite of the fact that RAF still felt Arty/R was not "on" owing to the risk unless fighter cover was provided'.[134]

Signal

From First Army

To War Office

Recd.0700 9 Dec 42,No.087611

RA Cipher 8 Dec

Most secret,From R.A.

Air op already an unquestionable success despite adverse conditions and serious local air inferiority. Besides Arty Observation has proved invaluable for liaison inter communication and contact patrol. One A/C shot down my Me.109 and burnt up. Pilot safe. Blind spots of high wing design and austers almost complete lack of view aft.seriously and unnecessarily endanger pilo. Follow on A/C must be low wing capable of carrying rear facing observer and of being assembled and dismantled quickly for carriage portee. Pilots have frequently obtained special or urgent information which could not be passed owing to lack of range of sqns.present high power number eleven sets.T.O.O. nil.

Copies to Director Military operations,Directorair,Director Signals,

General Lockhart in his despatches to India praised the AOP, he was given this briefing by Director of the Air War Office at the meeting of Dominion representatives on 23rd February 1943. 'Air OP has played a vital role in African Campaign'.[135]

655 AOP Squadron was raised at Old Sarum on 8th December 1942 under the command of Major E.B. Ballard equipped with Auster Mk 1 aircraft and carried out its initial training with army formations

134 AIR....GS Branch U.O 22366/RA1. dated 19th October 1943.

135 WO CGS 43/3 dated 2nd March 1943 from General Lockhart,India Office to Lt-Gen Morris C.G.S.

of southern command, by March 1943 it was equipped with Auster Mk 3 aircraft. In August 1943 the squadron now commanded by Major Oldman received orders to proceed to North Africa thus the squadron docked at Algiers on 25th August 1943 'sometime was spent in collecting equipment from Algiers and Bone and it was not until October that the squadron was able to commence training'[136] thus almost two months were wasted by a flying unit in just collecting its equipment and when it was able to conduct any worthwhile contribution in the war it was ordered to move to Italy as part of 8th Army.

Air War a Bird Eye View

Japan attacked American Pacific fleet on 7th December 1941,with the help of its carrier based airpower and sank eighteen warships, although all the three aircraft carriers luckily escaped the destruction. From this point onward the Japanese onslaught in Thailand, Philippine, Singapore, Burma was spearheaded by their respective army and naval air forces. The sinking of 'Prince of Wales' and 'Repulse' within hours by the thirty-four bombers and fifty-one torpedo bombers of Japanese naval air arm ushered airpower into a new dimension where naval power became subservient to it. Churchill wrote 'The efficiency of the Japanese in air warfare was at this time greatly under-estimated both by ourselves and by the Americans ... the superiority of their air force was decisive in crumbling the resistance of the British troops…in deciding the fate of Malaya and Singapore'.[137]

The Battle of Midway on 5th June 1942, 'one of the most decisive days of warfare in history,[138] by the battle's end the balance of WW2

136 Brief History of 655 Squadron,MAF.
137 Liddel Hart,History of WW2 {London,Cassell,1977} p 236
138 Peter and Dansnow, The Worlds Greatest Twentieth Century Battlefields{London,Random,2007} p 49

in the pacific would be irretrievably reversed, the decisive point of this Japan-American war was decided by the airpower of the America, when it sank the Japanese fleet, the first instance where a naval duel was not decided by ships itself.

Britain from the beginning of the war had no other means to strike back at Germany except through its airpower, which it did but with a heavy price and without any viable results. The pre war idea of winning a war purely with the help of a bomber force did not materialised. First the RAF did not had the means to deliver that and secondly the theme that bombers can defend themselves proved wrong and lastly the bombing was inaccurate. The Butt Report of August 1941 probed the effectiveness of bombing and came to conclusion that the total number of sorties that reached within five miles of their target is less than one third and 49% of bombs fell in open field between May 1940-May 1941.[139] The 22nd September 1941 bombing directive stated 'With a bomber force of 4000, they could destroy 43 German cities with a population of more than 1,00,000,[140] Chief of Air Staff Sir Charles Portal argued that with such a force the RAF 'could win the war within six months'.[141] The sum of experience since the out break of war had shown that the long established concepts of Air Staff and Bomber Command were badly in error.

On 14th February 1942,directive to Bomber Command emphasized 'Bombing campaign to be focussed on the morale of the enemy civil population and in particular of the industrial worker...Area bombing is to be preferred and targets are Essen, Cologne, Duisburg, Dussledorf as priority...you accordingly authorised to employ your forces without restriction'.[142] With this background of targeting cities and civil population along with erroneous bombing technique, on 22nd

139 Ibid p 623
140 Ibid p 624
141 Ibid 624
142 Ibid 625

February 1942 Air Marshall Harris took command of Bomber Command and bombing of cities became synonym with strategic bombing and strategic air power and this is generally what is remembered of second world war.

Harris started with bombing of Baltic port of Lubeck, Essen 8-9th March 1942 and reached its zenith with thousand bombers bombing Hamburg in July 24-27, 1943 which resulted in 40,000 deaths vast attacks on Berlin itself started in the winter of 1943-44. RAF was using incendiary bombs along with normal bombs to create firestorms. In this indiscriminate bombing, the whole cities were targeted although the [143]Casablanca conference{4 February 1943} and later allied Pointblank directives of 1943 laid the emphasis on precise military targeting. It was basing upon this directive that a strategic bombing in true sense was conducted against the Luftwaffe and German aircraft industry between March-July 1943 commonly known as 'Battle of Ruhr' the bombing accuracy had improved due to pathfinder techniques, the dams at Mohne and Eder were breached on the night of 16th May 1943.A total of 872 allied aircraft were lost in this battle. Yet there were no significant gains ,therefore allied again resorted back to terror bombing.

Battle of Ruhr was followed by Battle of Hamburg, in which thirty–three major attacks were conducted on that city and others between July and November 1943 involving 17000 bomber sorties. The only strategic target that was hit in this period was on 17th August 1943 when 597 four engine bombers hit Peenemunde research site of flying bombs 'The effects were not so great as was imagined in London'.[144]

Battle of Hamburg was followed by Battle of Berlin, lasting from November 1943-March 1944, it involved sixteen major raids on Berlin and twelve other targets were also hit 'Churchill encouraged it as

143 Ibid p 627
144 Ibid p 630

it pleased Stalin...20,000 sorties were flown...Germany was not brought to her knees nor Berlin...morale of Bomber Command was shaken by losses and it was abandoned'.[145]

Americans were also busy in bombing the Japanese cities in similar manner. The first such raid took place on 18th April 1942 when sixteen B-25 bombers attacked Tokyo without causing any major damage, but it was the only way of retaliating.

Sicily. British CIGS Alan Brooke was able to convince the Americans that next step instead of invasion of France should be the invasion of Italy[146],allies agreed in January 1943 on this but were able to finalise the plans in May 1943.Two allied invasion forces{three American ,three British and one Canadian division} went ashore in South East Italy on 10th July on 26 different beaches opposed by nine Italian and two German Panzer divisions. In July 1943 the invasion of Sicily took place .The whole question of AOP operating in mountainous terrain was put to test by Captain Nielson who visited the 52nd Division Headquarters and informally informed the divisional commander General Ritchie that 'AOP is not fit for operations in mountainous country basing upon the experience of Tunisian campaign'.[147] Nielson further highlighted that Captain Rogers an AOP pilot had crashed and died in the Wales because of intensive bumpiness and squadron commander himself advocated the restricted flying in mountains. General Ritchie in his letter to War Office stated 'it is unnecessary to consider...Air OP in mountain warfare any further...helicopter type of aircraft might meet our needs'[148] It became a big issue with War Office demanding the explanation, it resulted in a court of inquiry, it was finally the RAF which came to the help of pilot 'I feel that 52 Division are the author of their own mistake'[149] This episode

145 Ibid p 630
146 Simon Ball,p-183
147 AIR 39/143;wo/arty/6088{air 1}dated 17th April 1943
148 Ibid.Letter No.LDS/628/i/G/I,30th March 1943.
149 AIR 39/143.

highlights the basic difference between the approach of two services as far as flying was concerned. War Office admitted that the 'value of {AOP}squadron under mountain warfare conditions can only be gauged after trials have been carried out'.[150] One advantage of this episode was the evaluation of the AOP in which it was emphasised that AOP requires over '300 yards of flat smooth ground with in 2-3 miles of artillery position and flight duration should not be more than 20 minutes'.[151]

The first AOP segment consisting of one flight each from 651{three aircraft} and 654 squadrons arrived at Sicily on D+10 in a ship ,it virtually faced the same problems as they faced in Africa. The advance party of 651 Squadron consisting of six officers and 24 other ranks landed on D plus nine, the aircraft arrived in crates, It has been noted that in both the campaigns the aircraft were transported by road it was not the most viable means of transportation during the Sicily invasion the Auster were moved by road for over 60 miles in the back of a three ton lorry the bad state of the road resultantly caused damage to two aircraft 'I would stress again that this form of transportation cannot be done very frequently'.[152] The operations in the area were over by the time Air OP assembled its aircraft they had lasted till August 4th when the 13th Corps advanced in the south of Catania the next phase started when the corps advanced north of Catania into the foothills as far west as Mount Etna, thus squadron joined the action in this phase. 654 Squadron also arrived on 4th August they had the similar story, their 16 Auster were carted in the three tonners {five extra three tonners were provided by the corps}which travelled 45 miles over bad and winding roads but they suffered no damage, fuselage was put into the vehicle nose first with tail planes and rudders detached the fuselage being held secured by an iron cradle, the main planes were lightly crated in wooden frames and rested

---

150 Ibid,letter from major general DO Staff duties to GOC 52{L}Division dated 20th April 1943.

151 Ibid,reply of Major Cobley the 652 Squadron Commander in due course of court of inquiry.

152 Operational report of 651 Squadron from 19th July 1943 -15th August 1943 signed by Major Neathercoat.

on the seats along the sides of the vehicles latched.

651 Squadron was fully operational by 6th August in support of 13 Corps 'ALG's were at times 15 miles away from the affiliated regiment'.[153] This distance involved almost over an hour of drive thus valuable time was lost which negated the philosophy of AOP. The terrain was mountainous which made the preparation of the landing grounds a highly skilled affair and put strain on the resources. 'throughout operations ALG's have been very difficult to find…no observation up to the foothills, although AOP could obtain a certain amount of observation into these hills it was not practicable to fly out far into the plain'.[154] The field elevation also restricted the power out put of the Auster aircraft. Sicilian campaign as analysed from the 654 squadron war diary reveals that ,almost one hundred and forty five hours were flown by the squadron and conducted twenty nine sorties for artillery and flew seventy hours on sorties not related to artillery. Majority of sorties were undertaken for a duration under half an hour, almost 90% shoots were undertaken through the medium regiments, on ten occasions between 1st and 10th August, more than fifty rounds were fired by the artillery on the orders of AOP without causing any damage. B Flight flew almost 220 minutes on 15th and 16th August and did conduct registration, only once i.e. On 1st August 1943 more than 24 rounds were fired otherwise it was a solitary affair. One key feature of the squadron war efforts resulted in the neutralisation of a hostile gun area but it was conducted through clock code system, the very system against which the artillery was so against and which had been pleaded by the RAF for at least giving it a try. Squadron was useful in carrying out the contact patrol which it send after every two hours the aim of these contact patrols were to gather information. The tasks of the AOP in Sicily was mainly in the role of spotting and engagement

---

153 WO Major Neathercoat Operations carried by 651 Squadron in Sicily from 19th July -13th August 1943.

154 WO… Major T.Willett Squadron Commander 654 ,20th August 1943.

of hostile artillery batteries, mortars and Nubelwerfers{kind of mortar} normal registration and destructive shoots,reconnaissance of gun areas and information sorties.

Italy. In the latter phase of the campaign that involves invasion of mainland Italy, the Salerno {south of Naples} landing took place on 9th September 1943 with three allied divisions comprising of British 46th & 56th Divisions, Rangers to capture the passes that separates Salerno from Naples, Southern attack force consists of American 36th Division, only one German Panzer Division the 16th was on the coast. On 9th September 1943 one flight of 654 Squadron arrived at Tuschiana {Salerno Bay} and operated in support of 10 Corps and the American 5th Army. Squadron was completed in a fortnight and attributed its lack of activities to the terrain 'mountainous country north of Salerno was not so suitable for the Air OP's to operate.'[155] This squadron remained in Italy till the end of the war and is an example of AOP operating in the mountainous region. Its three flights kept on shuffling between the 10 Corps, the Canadian Corps, the Polish and Indian Divisions apart from working with the American Army. Seldom it operated as a single entity, it is a study of another aspect of AOP ,the comparison between the squadron operating as a centralised or in flights. AOPs activities fluctuated with weather and air situation, when the Luftwaffe was busy against the Anzio landing the squadron was able to fly at all altitude above three thousand feet 'free from enemy air...squadron was able to fly sorties lasting over an hour'.[156] The glaring difference between the AOP and American AOP[157] became obvious. American AOP were the first to land on the soil, four American pipers{AOP} cubs took off from the specially designed strip from a landing craft and directed the fire of their naval ships and later landed on the occupied ground, American AOP also flew over 2000

155 MAF,Short Historical Notes No.654 AOP Squadron.,p.2.
156 Ibid,p.3.
157 MR/192 report of the general board.pp.5-15.AIR 23/2552 letter to HQ air Command SE Asia dated 24th November 1944.

sorties in African campaign as compared to mere 37 by the British AOP, the probable reason was the unifying control of the aircraft and a basic difference in the tactics of both. The Americans kept all their aircraft concentrated at an airfield close to the front and all missions were taken and co-ordinated by the G staff where as in British AOP the aircraft were allotted in singles to brigades with no reserve and in the end it was the field artillery which suffered.

655 Squadron's aircraft flew from North Africa by way of Pantellaria and Sicily to Naples where they were joined by their sea party and on 21st December 1943 the complete squadron left Naples moved to a strip south of the River Sangro, by Christmas day placed under command 13th Corps of 8th Army taking over the duties from 651 Squadron. AOP moving to Paglieta on 2nd January 1944,A Flight of 655 Squadron was detached and placed in support of 5th Army to supplement them for Anzio operation south of Rome another flight of the squadron C Flight also joined in on 20th February. During Operation Honker which was launched against Cassino on 11th May the squadron did carried out some sorties but nothing worth mentioning in terms of artillery shoots but in terms of intelligence one sortie stands out which was flown by Captain Fortnum of C Flight on 25th May 1944 when he was able to locate troops 'I was astonished to see large convoy of armoured vehicles that appeared stationary…there was a distance of 800 yards between them and own troops… I had no means of communication with them so I landed there and found them to be American…they reported that coast was clear of enemy forces…I called my regiment which duly passed the good news to VI Corps…I was actually the first man from the beachhead forces to contact the army in the south'.[158] 655 Squadron lost five pilots in the campaign the highest number of casualties by any AOP squadron, one pilot captain Peter was shot down by the Luftwaffe patrol, Tony met

158 Squadron history 655 Squadron,Air Historical Branch{A.H.B.5}December 1962,MAF.

the same fate, John was shot down by the ground fire, Aitchison was shot by own fire. B Flight supported 6 AGRA[159] and that was the only time that an artillery group was supported by the AOP effectively, it had four medium and one heavy regiment. Following the capture of Rome in June the squadron was concentrated at Santa Maria La Fossa for a period of rest and training during which it received its first Auster Mk IV aircraft. Squadron later resumed operations under 13th Corps in July and replaced 657 Squadron, one high mark for the squadron and for the Air OP was the visit of the King who was flown in an Auster to Maligano on his inspection of Italian Forces. In August 1944 the squadron supported the 13th Corps in its operations including the advance to the River Arno and the capture of Florence, in August alone squadron flew a total of 651 sorties, there was a static warfare during the winter season and only the launching of the spring campaign of 8th Army resulted in some operational activity for the squadron when it moved to Lugo area where its B Flight supported 10th Indian Division. With the end of war in Italy the squadron was employed in information role carrying out sorties to report on enemy movements, dropping propaganda leaflets and acting as ferry flight especially in Milan area. Squadron was disbanded on 1st August 1945.

One key handicap in evaluating the effectiveness of the AOP is their over exuberant reports and returns which does not match with the instructions of Air Ministry and War Office on the subject, the common factor in majority of the reports are praises for the artillery staff of the corps 'through out the operations we have been most intelligently employed and have received every co-operation from RA 30 Corps and divisional artillery and have been used with discretion and regard for our limitation...shooting procedure presented no problem'.[160] The aspect of night flying was another issue, despite the propagated capability of AOP

159 MAF. Box No.10 Lieutenant Colonel Oldman,History of 655 Squadron.

160 WO ;report on the activities of 654 squadron in sicily

the Air Ministry highlighted that 'pilots of 655 squadron got blind while landing at night using glares'.[161] Ministry further clarified the night flying capability by stating that it can be done if moon is in second phase, Air OP should not fly more than 20,000 yards into enemy territory, remain within wireless range and operate from same strip.[162] 'Air Op Squadron was operating from D Plus 1 day and was an unqualified success...no aircraft were lost through enemy actions...Air Ops also observed many naval shoots...Auster aircraft gave excellent results'.[163]

Table 1 Analysis of Sorties No.654 AOP Squadron, October 1943-April 1945. In the Italian Front excluding Sicily campaign which ended on 17th August 1943 and invasion of Italy mainland started on 3rd September 1943.

1943	Effective shoots	Non effective shoots	Counter battery rec-conaissance	Informa-tion	Total
October	136	26	33	115	310
November.Weather Bad	53	24	8	37	122
December.Bad weather	31	23	28	46	128
1944					
January	136	46	90	171	443 309.25 opera-tional hours 155 non operational hours
February Cassino Battle	180	40	112	25	357
March-Anzio	79	26	203	28	336
April	112	7	36	13	68
May,Adolf Hitler Line 13 Corps,1 Canadian corps	162	34	246	94	536
June.10 Corps	128	15	207	44	394

161 Ibid
162 AIR 23/2552 :CS 11016/111/D.Ops{tac}dated 7th November 1943
163 AIR.... U.o.22366/RA.1 dated 19th October 1943.

July	509	41	403	28	981
August,Gothic Line	455	73	274	25	827
September, PolishCorps	472	70	185	24	751
October, bad weather	244	30	117	5	396
November ,bad weather	137	42	144	20	343
December, bad weather	112	Not Re-corded	125		237
1945					
January bad weather	126	Not Re-corded	142		268
February	228	NR	183		411
March	101	NR	86		187
April,Senio River	592	NR	391		993

Chapter Four 1943-1945

Mainland Europe & 652 Squadron

In June 1943 UK established a Tactical Air Force located[164] at RAF station Bracknell in accommodation previously occupied by Army Co-operation Command, which ceased to exist and was disbanded on 1st june 1943. The AOP Squadrons continued to remain under command the ex Army Co-operation Wings associated with the armies to which the squadrons were allocated, it was highlighted that at later date these squadrons will be transferred to the appropriate composite groups as part of B.E.F.

When and where the great landing will take place was probably the best guarded secret of all time. Shortly before D-Day the squadron commanders received two separate envelopes each marked as Top Secret, 'on opening each of them I found another envelope which was marked Top Secret ...the message in the first envelope was "signal estimated requirements of boiled sweets" the message in other envelope asked for the requirement of prefabricated crosses'.[165]

Employment of AOP in Operation Overlord from a historical perspective is similar to the RFC's role in early part of the WW1. AOP pilots were flying the similar un- armed aircraft which had the same characteristic as in WW1 in terms of flying envelope, however in this war the role was more specific than the past; to act as an elevated platform for the artillery. The meagre available publish material leaves a vacuum

---

164 WO 233/60 (s/06/173/ops dated 15-6-73)

165 Andrew Lyell,p-72.

as to the real evaluation of the AOP in the light of previous almost six years of hard paper battles between the army and the RAF where both had contradicting angle of evaluation of the concept. The key element from regimental history is the admission of German Commanders on the menace of AOP 'but the greatest menace of all are the slow–flying artillery spotters which work with utter calmness over our positions, just out of reach and direct artillery fire on our forward positions'.[166] Yet these historian have very conveniently over looked the fact that over 1300 American Air OP's were also operating in the same area, thus remarks of German commanders are not meant for only British AOP. 21 Army Group had No's 652, 653, 658, 659, 660, 661 and 662. AOP Squadrons comprising of over one hundred Auster IV and V aircraft under its command, this army group had since 1943 contemplated the use of AOP in the assault phase similar to the American use during the Anzio, 652 Squadron did carry out few deck landings but the concept was not adopted mainly as it compromised security. It was at this stage that 21 Army Group requested War Office to procure American aircraft but former regretted that it is not possible at this stage.

AOP Squadrons prepared hard for the incoming invasion and conducted flying over the water and remained with the affiliated formations. Tactically AOP flights were affiliated with the divisions and as such it was planned that advance parties from each squadron should get ashore following the invading infantry, establish the ALG which were selected from aerial photography, level and clear the strips from obstructions sufficiently enough for safe landings and most importantly to establish a signal network to inform the Flights waiting in UK, thus Flight Commander's party was split into two, flight commander leading on his flat feet and another led by a pilot bringing in the two 3 tonners which were the only transport that flight was allowed in the first wave.

---

166 Parham and Belfield *Unarmed into Battle*,p,73;Major Andrew lyell *Memoirs of AOP*,p.78

These parties were also task to establish liaison with the formation by having a representative preferably an officer attached with them, for this purpose wireless set No.22 accompanied. This was an utter wastage of pilots as in a flying unit it is the pilot which matters, it takes years to make a pilot and AOP was sending them in ground role. Unlike RFC the AOP had no centralised command to plan their use.

As all these AOP squadrons were raised in same time period and had the same aircraft thus for the purpose of evaluation it is logical to follow one squadron's activities which holds true for others.{published works have been portraying broader picture thus this aspect has been neglected}. 652 was the first squadron which trained with Royal Navy for cross channel crossing exercises in November 1943 and thus conducted operation from deck ' besides being pioneers of deck landings the 652 was also the pioneer of air photography from Austers also experimenting with night photography and night shoots'.[167]

652 Squadron in support of 1st British Corps had its advance party A Flight landed on Queen Sector {red and white beaches} at 0045 hours on 7th June 1944.[168] The general framework of the assault was split into four waves, assault, 60%, 75% and war establishment in terms of manpower and equipment, thus one 3 ton lorry per section of flight as per its original table of equipment and organisation was more than that was allowed under the circumstances.

Squadron flights were supporting the corps formations in following manner.

SHQ.	Captain Nicholson, 1st Corps Headquarters.
A Flight.	Captain Loveridge, 3rd Canadian Division.
B Flight.	Captain Neilson, 3rd British Division {7th, 33rd and 76th

167 MD 2878, *Short History of 652 Squadron*, MAF.
168 MD/4024 Creswell, *Collected pilot note books B Flight AOP Squadron*, p2

Field Regiments}

C Flight Captain Vipond, 51st Highland Division.

D Flight Captain Murray, 8th AGRA{this flight was attached from 658 Squadron}

ALG's were selected from air photograph but situation on ground 'fog of war' was different, A Flight landing ground was under close enemy threat[169] thus another strip was selected at Beny-sur-Mer, A Flight became involved in fighting with the Germans and one soldier was wounded, the flight was still busy in fighting when the next wave early morning brought B Flight on shore. The layout of the B Flight was as under.

Captain Nielson with the assault marching party, Captain Keen with two 3 ton Bedford troop carrier vehicles carrying the initial equipment, one Jones 150 cc motor cycle, spares, petrol, lubrication and LeRiche electrical flares, Captain Linton with signallers and wireless set was the liaison officer at the 3rd Division; the wireless set was on a hand cart. The three remaining pilots of the flight were part of air party, 60% vehicle and personnel party, 75% vehicle and personnel party and finally war establishment party led in by the squadron adjutant.[170]

Three officers {including Flight Commander Captain Neilson} of B Flight after getting on ashore set out to locate the ALG, Captain Linton went to contact the division at Collevittle-sur-Orne and was duly informed that intended ALG at Cresserons-Plummetot is still under Germans control thus they probed for alternate ALG resultantly the arrival of air party was delayed for a day. On **D+2**, air party took off from Old Sarum at 0700 landing at 0815 hours ,on this day only five aircraft supported the two divisions the 3rd British and Canadian, squadron

---

169 MD/2878, *Short History of 652 Squadron*.also see *Brown with two shades of Blue*,p-5.

170 MD/2878 Ian Neilson,*Transcript relating to the role of the AOP in Normandy 1944*Preparations and organisations Assault Normandy.

suffered its first fatality when Captain Eric Puch of A Flight was shot down by the Luftwaffe 'this was probably due in no small way to the conspicuous black and white marking on upper surface wing imposed on all allied aircraft'.[171] No 658 squadron tale is slightly different as although its three flights went in similar fashion its commanding officer lost his way and took almost two and half hours of flying to reach the ALG from Selsey to Normandy 'landing ground was a flat piece of ground rolled by the advance party three ton truck...no 'T' to show the ground...well camouflaged...I landed at other field'.[172]

652 Squadron operated from a strip which was large irregular shaped field on the eastern boundary was the anti aircraft command ground control interception station and tented camp, on other two sides a small river and lightly wooded area.

Captain Orr was an expert in photography and by12th June the whole of 1st Corps area had been photographed. On 13th June Lieutenant Vann of C Flight along with his observer were killed when they were shot down by hostile small arms fire. 23rd June was a particular day when the squadron provided continuous cover over the entire day, a total of 57 sorties were flown in support of attack by 51st Highland Division on St Honorine La Chardonnerette, the squadron pilots were able to engage 20 hostile batteries one pilot locating seven during a single sortie.

Squadron adopted the tactical doctrine of keeping all the flights centralised rather than attaching them with regiments, one cannot pass a comment because commander on ground is the best judge but it was contradictory to the very principal of Air OP 'commanding officer and headquarters continued to interfere in the operations of the flights... results were mixed'[173] it required a pilot to be stationed with the divisional

171 Ibid.

172 Andrew Lyell, Memoirs of an AOP officer.pp.72-73

173 MD/2878 Ian Neilson, *Transcript relating to the role of the AOP in Normandy 1944.* Major Nielson's papers and written account, the written account is part of squadron history at MAF.

headquarters for liaison but resultantly the squadron was deprived of a flyer.

11th June 1944-15th August 1944, in this period the 1st Corps were ashore and lasted till the days of Falaise Pocket,on the east bank of Orne river were the 6th Airborne Division controlling the open ground towards the Bois de Bavent, northwards to the mouth of the river opposite to Ouistreham and southwards to the divisional boundary with 51st Highland Division which was the corps reserves, away on the right was the 30th Corps. B Flight supported 3rd British Division which had 7th, 33rd and 76th field regiments all equipped with 105 mm self propelled guns the heavy and medium guns were deployed in the rear and were available for harassing and counter bombardment fire all this artillery was under 4th AGRA. The maps issued and used by the pilots were 50,000 scale they were quite useful as long as the target was on or near some definite landmarks. By 12th June all the aircraft were housed in bulldozed pitts,on the same night the landing ground came under Luftwaffe attack and a 500 pound bomb hit the squadron area but no major damage was done. Squadron and flights were well dispersed and adequate slit trenches were made. Many Germans soldiers had kept dogs as pets and these dogs remained beside the bodies of their masters until they were buried and then ran wild,some were adopted by British units.A Flight acquired an adorable spaniel.

During the battle for the Caen which lasted till 10th July A Flight supported 2nd Canadian Division and later 49th Infantry Division, during the Battle of Caen mass bombing was carried out by the Allies during which the pathfinder aircraft dropped markers over the Polish & Canadian formations and it was the squadron pilots which realised this potential catastrophic catalyst; they were able to stop the bombing on own troops by immediately communicating through air craft signals to own bombers.[174] AOP missions were also hindered by the bad

174 Ibid.

weather, the torrential rain of 20th July virtually made no flying possible for another four days. Low clouds and rain severely limited Air O.P's usefulness[175] by then squadron had flown 856 sorties totalling 420 flying hours including 458 shoots, loosing two pilots one rear observer and three aircraft in the bargain.

16th August-4th September 1944. After breakout from Falaise the AOP for the first time supported an armoured division. AOP was not organised for rapid movement as it lacked requisite transport for occupation of ALG's thus a liaison officer had to be detached and moved with the armoured division. Breakout from Falaise area to Seine by the allies cut the Germany's 7th Army, B Flight supported 7th Armoured Division till 4th September it moved to a new strip near St.Pierre Sur-Dives popularly known as Bel Air 'it smelled strongly of dead horses and good Germans',[176] it also took part during the siege battle at Le Havre. The armour was swift in movement and battle became fluid it was almost an impossible task to engage targets without causing friendly damage to own troops. Flight had to move constantly, the AOP was not designed organised or equipped for a moving battle, the representative at the division became mobile and attached with RHA regiment within the battle group that moved forward in front of the main division, his role was to receive briefing for sorties, collection and collation of information regarding own troops and the enemy in addition to this he had to ascertain the areas that were good enough for landing strips, he has to pass all this information to the flight on wireless which was not effective 'in fact the armour was moving at such pace that we were out of communication range...Air OP was not in exceptionally great demand'.[177] One key feature of this phase of battle was the vulnerability of the unarmed aircraft especially to the German flak, the armour division

175 Ibid also see squadron history.
176 Ibid.p-12.
177 Ibid.p-13.

pressed on the main road leaving the flanks until later, Captain Smith of B Flight was hit by the enemy flak just 100 yards away from the main road. It was at St.Pierre that flight's entire stock of petrol was destroyed in an attack by the Luftwaffe which used incendiary bombs.

B Flight later saw action at Dunkerque, Arnhem Corridor, Antwerp, and ,Breda before the crossing of Rhine. 'Between June 1944 -29th April 1945 it conducted 368 artillery shoots'.[178]

River Seine was crossed on 3rd September 1944 and squadron supported 1st Corps{49th & 51st Divisions} in its operations to capture LeHavre, now the sorties were mainly conducted to locate enemy flak guns with little interference from Luftwaffe, normally conducted at medium level, pilots used binoculars as well which indicates the decreasing resistance, night sorties were carried out but were not successful. 1st Corps was now in the rear of the main battle, 30th Corps having occupied the Antwerp a week earlier, on 23rd September squadron moved into Belgium with 1st Corps, A Flight was attached with 49th Division in the area of Turnhout-Antwerp Canal, C Flight supporting 1st Polish Armoured Division near Breda another flight A Flight of 661 Squadron was also attached commanded by Captain Courtney. On the day of moving to LeHavre the B Flight moved from Foucart to Angerville where 'CO of the squadron appeared suddenly out of blue performed a heavy landing, pranged the undercart of the of aircraft, ordered the flight to move to Vergetot took another aircraft and disaapeared again,after arrival at Vergetot fresh orders were received to move to Le Prunier this was accomplished by 2230 hours…four sets of orders and three landing strips in one day for one move must be an all time record'.[179] The LeHavre garrison surrendered on 12th September 1944. B Flight alone flew 52 sorties during which 36 hostile artillery batteries were located

178 Squadron History and Nielson's account.p.35.
179 Ibid.

and eight registrations and two neutralisations ranged and observed.

13th September 1944-22nd March 1945

Life became very quiet after the capture of LeHavre for the Air OP pilots and nothing other than occasional ferry trip were conducted the pilots now had the accommodation in a chateau where a whole day was wasted to get the water system working the squadron had its get together on 17th September as well, 1st Corps now concentrated in area Yvetot-Bolbec. B Flight was detached from the squadron and ordered to Dunkerque on 20th September, the containing force at Dunkerque was commanded by Brigadier Routledge, 48th RN Commando,150 and 191 field regiments apart from 109 HAA regiments were the fighting elements. One section under Captain Clarke was attached with 150 regiment at a strip near Le Pane; he was able to visit the regimental cinema when commanding officer of the regiment send his own soldiers for sentry duty at the strip.The anti air craft guns were used in ground role in this action

Squadron headquarters and B Flight in mid October were combined together for 1st Corps push towards Esschen. In December 1944 the last dice by the Germans were thrown in the Rundstedt's Ardennes offensive. A Flight of 652 Squadron supported the 6th Airborne Division & 4th AGRA while operating from Mettet and Dinant strips. Weather played a key role as the fog sets in at the end of the month with alternating high winds and rain thus continuous cover was not possible ,the move of the A Flight's entry into the battle was initially delayed for few days because of the weather. Squadron supported three operations in January 1945, Operation Trojan was aimed at simulating large scale crossing of the River Maas mainly to force Germans to reveal their strength,C Flight supported the 1st Polish Armoured Division in this regard simultaneously B Flight also carried out smoke and counter mortar sorties. Operation Horse was the second operation in which the commandos attacked an island{Kepelsche Veer} in the Maas,attack however was a failure.[180] Operation Elephant was more ambitious in nature launched by 4th Canadian Armoured Division and was supported by the squadron by providing cover continuously for four days, meanwhile A Flight supported 7th Armoured Division & 3rd AGRA during their attack on on the Julich-Recermond Salient, Captain Searle was able to get a direct hit on a German tank in one of the sortie which he celebrated with aerobatics and low flying exhibition and it was during this performance that he was attacked by the Luftwaffe and he landed with his aircraft riddled with bulletts and miles of telephone wire trailing, later. A Flight also assisted Dutch intelligence organisations in communicating with their companions in German occupied areas. C Flight took part in 30th Corps battle in Reichwald forest in February 1945 on the approaches towards the Rhine River supporting 51st Division after this battle the C Flight moved to Dunkirk in support of 17th AGRA and while its convoy was moving one V-2 rocket landed in the middle of it destroying one vehicle and killing the driver Bombardier Strauch; he was a good cook.

The artillery shoot procedure for which all this effort was undertaken, its effectiveness is of dubious nature as the artillery regiments refused to act on the pilots executive command of fire, one reason was the deviation of the AOP principles that pilot should be the battery commander, which was not possible here and neither it was ever practiced in the history of AOP. To bring fire quickly was the other key factor for the raising of these squadrons but the flights had very little idea of the actual location of the artillery regiments. In actual shoots the concept of ranging was seldom applied as it was time consuming in such

180 652 Squadron History,p-6.

hostile air environments, it was far cry from WW1 when fighters used to provide the air cover for the shoots. Now the pilot had to fly, locate the enemy and plot it on map, communicate to the guns, bring the fire of own guns, correct it and supposed to follow it by procedure and all this in a circular pattern of flying behind own firing lines. It was simply not possible at least in the moving battle, thus pilots resorted to call for the maximum fire 'Stonk' of every gun on the suspected target but this the ground gunners refused to do so on the reason of ammunition scale involved. During WW1 it was the observer's job to locate the German artillery but now the observer was only there to look out for the hostile aircraft and these observers were not trained in most of the case they were the technical airmen as they could differentiate the aircraft ,it severely affected the maintenance of aircraft as Auster had to be serviced after forty hours of flying. The reporting of AOP pilots was another cause of concern., in one case a pilot Captain Bawden of C Flight ex 652 Squadron 'spotted in open ground south of St. Honorine some thirty enemy tanks'.[181] The same sortie was task to the commanding officer of 658 Squadron Major Lyell who flew on 12th July 1944 'but I reported that what had been seen were only bushes'.[182]

AOPs were under command of the artillery and they were employed for the dawn to dusk observation of the front similar to the WW1 pattern of contact patrol but it was not useful here as the targets which AOP would select were not the priority for artillery, AOP would select what ever they observed and in most of the cases these were lone machine guns or mortars. Locating of enemy artillery guns was extremely difficult as 652 squadron history admits about an enemy gun position which have been shelling the squadron on regular basis 'in spite of many sorties to locate it remained un seen to the end'[183] another

181 MAF,A short History of No.652 AOP 1942-1955,part 1 {1942-45}p.3.
182 Andrew Lyell, Memoirs of an AOP officer,p.80.
183 MAF,A short History of No.652 AOP 1942-1955,part 1 {1942-45}p.3.

factor was the pilots lack of tactical knowledge of the front ,one of the key reason for raising of AOP. In one sortie Captain Hell's attempted to destroy a bridge through artillery fire only to learn later that this bridge was the life line of own infantry. The dust of the tanks and constant aerial bombing nearly blinded the pilots as a wall of dust was raised 'after the commencement of the battle it was impossible to see any thing owing to a smoke screen and dust'.[184] The only occasion in which AOP justified its existence in terms of artillery concentration fire was on 17th July 1944 when virtually artillery of entire 21 Army Group was placed under 658 Squadron for shoot this was an exception rather than a rule.[185]

Operation Plunder, Rhine Crossing.

The operation began at 1530 hours on 23rd March 1945 with an intense artillery bombardment also heavy smoke scree was used to mask the crossing thus Air OP had little to do other than to watch the proceedings from a grand stand, B Flight was supporting the 3rd British Division the holding division on the 30 Corps front through which 51st Division advanced to take Rees on the east bank of Rhine, the flight was able to carry out three sorties before the darkness on the same very day mainly for counter battery bombardment purpose. On the night of 23/24th March the amphibious assault was carried out and by first light both 51st and 15th Divisions were secure in their bridgeheads, airborne landings were carried out on the morning of 24th March the visibility was so poor that nothing of value could be seen. Germans still held on to the Halden which was bypassed by the 51st Division, the aircraft at strips had to face a new menace; allies gliders passed over them in one case one glider broke into two and felt on the strip, the ground crew also rescued crashed

---

184 MAF ,*War Diary of 658 Squadron ,entry for 15th July 1944*, Andrew Lyell, *Memoirs of an AOP officer*,p.81.

185 Parham and Belfield *Unarmed into Battle*,p,78;Major Andrew lyell *Memoirs of AOP*,pp.82-85

Lancaster crew which came on the strip. AOP by and large carried sporadic shoots, by 28th March divisions were cleared of the bridgehead areas and B Flight moved up in support of 3rd Division advancing on axis running northeast through Enschede to Lingen on the Dortmund-Ems canal the flight was now placed under 662 Air OP Squadron,on 29th flight moved to Rees strip from where two sections were attached to support 76th Field Regiment. On the other hand A Flight supported 1st Polish Armoured Division after the Rhine crossing advancing rapidly through Razel, Emmen, Aschender and Upjever to Wilhelmshaven, landing grounds were difficult to find several aircraft were damaged while landing in this boggy soft country. Squadron headquarters was engaged in these final moments of the war in Deventer-Appeldon area.

By 3rd April 1945 the 3rd Division had reached Dortmund-Ems Canal at Lingen and were in a process of forcing the canal crossing, the B Flight also moved to a strip near by on 4th April but it became unserviceable after two days due to soggy nature of the soil and was also strafed by the Luftwaffe. The last evening of the war in Europe was inadvertently spent by all pilots on the landing strips.

A brief summary of B Flight's sorties flown between 6th June 1944 and 29th April 1945 highlights the overall spectrum of Air OP in the war. Operational Scrties including operational moves and reconnaissance forward for new advance landing ground. 1066. Domestic sorties including ferry and communication duties were 108 and effective shoots were 368.

American Air Observation Post

The Air Corps Tactical School was first opened at Langley Field Virginia in 1920 and designated as Air Services Tactical School in 1922, in 1931 school was shifted Langley to Maxwell Field Alabama it had the motto of Proficimus More Irretent{we make progress unhindered by customs}. American Artillery had the school for artillery training established in April 1824 at Fort Monroe where newly commissioned officers received one year training, infantry and cavalry schools were formed later.

Total field artillery aircraft operational in the European Theatre on 8th May 1945 numbered 1380 including L-4 Piper Cub and L-5 Stinson Sentinel. Each division had contained 10 organic aircraft, each corps had 30-70 organic aircraft depending upon the amount of artillery it have and each field army had 200-300 aircraft.

The organic Air OP's of divisional field artillery operated from a joint base in a centralised manner this resulted in battalion commanders loosing tactical control of their aircraft which initially they had, the aircraft operated on a schedule manner there by catering the battalions requirement, the aircraft were controlled by the divisional artillery even those battalions that were attached with the division pooled their aircraft for operation and operated from base strip. The primary mission of the Air OP aircraft remained the location of targets and the adjustment of artillery fire and to achieve this end the American adopted a technique of continuous dawn-dusk patrolling of the area there by having a direct daylight observation of the enemy area as long as the weather permitting, it resulted in reduce artillery bombardment from the Germans 'many times the enemy will not fire his artillery while we are in the air, however when making a turn we have noticed gun flashes and have been able to locate them, enemy evidently believed we are blind during a turn'[186] to counter this the Germans put their guns more in defiles there by forcing the aircraft to fly high. American Army employed the method

---

186 282 Field artillery Battalion, A report on American Air OP in European theatre, Fire Power Museum.p-5

of forward observer for adjusting the field artillery fire, sensing's were expressed in terms of yards right or left and over or short of target, Air OP referred smoke or white phosphorus for the first few rounds during the adjustment to prevent loss of initial rounds. 5th Infantry Division highlights the results of Air OP during the month of September 1944 as follow, Enemy guns destroyed were10,enemy batteries located were 61and enemy batteries neutralised 50. The American V Corps adopted procedure under which any pilot-observer could adjust the fire of either division or corps artillery battalion located within the corps zone.[187] The extent to which Air OP's were employed by the Americans in the Europe is highlighted from these figures. First Army had 16,110 fire missions and 62201 total missions were flown, Third Army had 22972 fire missions and 87569 total missions flown least were flown by the 15th Army with only 334 fire missions out of an over all missions of 5732, overall the American Army flew a grand total of 2,44,732 missions/sorties out of which 62,640 were flown for fire.[188]

Americans historians are candid to accept that adjustment of artillery fire of heavy artillery was not a satisfactory mission for Air OP mainly because targets could not be identified while flying above own area and any venture into the enemy area required an air cover for the Air OP. In case of adjustment of mortar fire the observation of Commanding General 29th Infantry Division states 'success has been obtained in adjusting mortar fire by air observation using the L-4H type of aircraft, the observer is assigned a target selected by infantry, the mortars are laid and an initial round of smoke fired and from this point the air observer conducts the fire...50 right, 200 up is the command for the mortars to fire 50 mils right and 200 yards over the last round observed'.[189]

187 Operational Memorandum No.11 ,Headquarters V Corps Artillery dated 3rd August 1944,Fire-Power Museum.

188 It includes 1st ,3rd ,7th ,9th and 15th US Army.

189 A report on American Air OP in European theatre, Fire Power Museum,p-7.

Briefing of an Air OP is the fundamental key for the success of any sortie, American Divisions had a somewhat different lay out from the British, they had S-1, S-2, S-3 and S-4 staff officers in the division, taking example of 28th Division Artillery which had similar layout, the senior air observer was the S-2 whose duty was to gather all intelligence data, maintaining an operation map and briefing pilots prior to each flight he also provided photographic data to pilots. In 133rd Field Artillery Group a system was evolved 'the commanding officer of this headquarters has instituted a plan which has shown excellent results for a period of one month's operation...the command post tent is installed on the field complete with stove and is used as an immediate headquarters for the pilots and observers, inside the tent is a bulletin board on which are kept an operation map and all latest air intelligence data, pilots and observers inspect the board and map before each flight and return after each flight to leave a written report of their observation'.[190] The landing strips were maintained by the army, corps and divisions in their area of responsibility, local security of the strip was at times augmented by the use of infantry but generally the guard was from artillery. American Army realised the importance of observer more than the British Artillery ,the observer job was to locate the target and adjust artillery fire thus for all practically purposes the Americans adopted the first world war system, it was on 12th September 1944 that commanding general of 12th Army Group authorised one field artillery observer on flight status per take off pilot however no additional personnel were authorised in this role other than the men posted in the regiment. British Air OP suffered mainly because of lack of communication and to counter this the Americans had special nets, a series 600 radio set was located near the very high frequency radio that communicates with the pilot, use of wire communication was quite extensive in passing the information to the bases from headquarters. SCR-610 set that was used in the aircraft was found to be highly effective

190 Ibid.

for the air-to-ground communications but the major drawback of the set was its weight for safe mounting in L-4 aircraft.

Each field army maintained a replacement aircraft pool with its army air force maintenance unit and it were the pilots of this army air force which ferry and delivered the aircraft where ever they were required, air craft were listed by their serial numbers and were generally delivered on verbal orders with paper work following.

American losses[191] in the European theatre of war taking example of 3rd Army for the period between 1st August 1944-8th May 1945 are as under, pilot error 49.89%{178 losses}operational accidents 17.19%{61 in number} combat losses 33.1%{118 in number} total losses were 357 in 3rd Army, average loss per thousand hours of operational flying was 3.79 and destroyed or salvaged per thousand hours of operatio comes to 1.54.

191 The complete figures were not available at the compilation of report,p-29

Map Courtesy **"FIRE BY ORDER"**. By **E.W.MASLEN-JONES**

Chapter Five

Operations in Burma

In February 1942 the Japanese 15th Army invaded Burma from Siam in the east and by March the British forces were on the run. Burma had been a part of British India and was separated from it in 1939, it had a frontier of over 700 miles with India, running north from the Bay of Bengal to the edges of Himalayas touching China. Burma is a green jungle country 'trackless, disease infected, unmapped in certain areas, world's worst country, breeding the world's worst diseases and for six months world's worst climate[monsoon }'.[192] The nearest principal town was Calcutta the main seaport almost 800 miles in the rear of the front.On completion of retreat the army took up defence along a line of between the River Chindwin and Brahmaputra.

By November 1943 Field Marshall Slim was commanding the 14th Army with almost 18 infantry divisions under his command it was the old Eastern Army of India. In August 1943 the British and Americans formed a new set up South East Asia Allied Command{SEAAC}to control all forces in Burma, Ceylon, Malaya, Dutch East Indies, Siam, India & China, Admiral Mountbatten was the supreme commander, General Giffard as C-in-C 11th Army Group. Air Marshall Pierce was the Allied Air Commander in Chief, Major General Stratemeyer was the US commander, Eastern Air Command had the 3rd Tactical Air Force under Air Marshall Baldwin within which RAF had the 221 & 224 Groups. No.22 was in Ceylon a new group 224 was raised, new advanced Air headquarters was established at Calcutta. In March 1942, 50 million Pound Sterlings were sanctioned for construction of airfields till then

192 Field Marshall William Slim *Defeat into Victory*[ ],p-169.

only 16 all weather strips existed in India and India did not had a single Radar station; Burma had one. The allies' suffering from malaria & dysentery alone were 2,50,000 in the year 1943/44.

Operations in Burma were different from the Europe and Africa, the very first factor was the distance away from War Office and Air Ministry secondly AOP here was not part of any amphibious operation. And last but not the least was the fact that only one squadron No. 656 was supporting the entire 14th Army and its three corps in a jungle terrain where the inter formation communication was lateral interspersed by the high passes, poor primitive road, rail and river communication, variation in the altitude had severe impact on the aircraft performance, on one end was the Bay of Bengal and at the other over five thousand feet high Imphal and Kohima. Similarities to previous campaigns remained the same notably the tussle between the army and RAF for control and command of the lone squadron, artillery on regular basis raised doubts about it and repeatedly the same answer was given by the Air Ministry 'OP are RAF units operationally under command to the army'.[193] By the end of war in Asia three AOP squadrons were operating in India the No 656,658 and 659. The last two squadrons arrived when war was almost over thus it is the 656 Squadron which is of interest. Available published literature is a lone book of memoirs written by a AOP pilot, squadron war diary is another source, the impression that arises is epitomised by General Sir Martin Farndale.

'Their flying hours broke all records, their feats of endurance and airmanship in appalling weather conditions set standards seldom equalled and they never once failed to answer a call for help from their comrades on the ground...story of remarkable feats of arms and must be recorded for ever in the annals of the British Army. the Royal Air Force and Royal Artillery.'.[194]

656 Squadron arrived at Bombay in August 1943 however its aircraft the Auster-3 arrived in December. AOP Squadron was placed

---

193 AIR letter No4500/1/Org.4 dated 27th August 1943

194 Maslen Jones, *Fire By Order*, back page cover.

under the No 222 RAF Group and on army side was supporting the 14th Army which had three corps 33,31 and 15 Corps. AOP squadron was not placed under the G staff but remained under the artillery staff.

The convoy carrying the pilots of 656 Squadron was diverted enroute and as such they arrived much earlier than they were expected. The pilots were initially accommodated in Juhu area of Bombay close to RAF transit camp. 'This was so remote from the concept of going to war that it was hard to believe…we had comfortable living, bearers to look after us…fighting was taking place 1500 miles away to the east … life in this corner of the Empire was virtually unaffected'.[195] Two weeks later squadron was move to Deolali where artillery also exist along with an airfield. Six Tiger Moth aircraft were given on temporary loan by the Royal Indian Air Force. The squadron Austers arrived at Bombay in December 1944 with another shipment at Calcutta; thus mechanics first assemble the Austers at Bomaby and later had to travel 1500 miles to assemble others at Calcutta. This all happened because of different markings on the crates.

In November the squadron officers had an area familiarisation tour visiting 4 and 15 Corps areas by air, road and boat. The conclusions were that squadron can provide support to only one corps either 4{Imphal} or 15{Arakan}, and it is more easy to do so with the 15 Corps mainly due to the poor power limitation of Auster at altitude of 5000 feet in case of 4 Corps area. It was also recommended that flights would not occupy the advance landing grounds at night and would retreat to rear.[196] In the end squadron headquarters and two flights{A &C}were attached with 15 Corps and B Flight remained at Johu for a planned amphibious operation which was later abandoned.

A Flight established itself at a suitable landing at Chota Maugnama

195 Fire by Order,p-9.

196 AIR/ 339;Report of members of 656 squadron{AOP}Squadron to 14 Army,30th October 1943.

in support of 5th Indian Division and C Flight in support of 7th Indian Division at Taung Bazar, 15th Corps Headquarters were at Bawli Bazar. On the Japanese side General Hamaya was commanding the 55th Division. On 21st January 1944 Flight took five hours to fly from Barrackpore to Chota Maugnama with refueling at Jessore, Dacca, Ferni and Chittagong.

The very first operational sortie of the squadron was carried out by Captain Frank ;the 5th division had just launched an attack and he was asked to carry out a reconnaissance sortie, 'he saw clearly that attack has met stiff resistance, he saw clearly that forward tanks and infantry were pinned down, he could also locate the principal enemy positions'.[197]

General Hamaya made the opening moves and within hours had encircled the 7th division and by midnight cutoff the airstrip from the divisional headquarters 'this was to be the most severe baptism of fire for all of us'.[198]

RAF men of the Air OP were least prepared for this situation, it was frightening 'each night the Japanese were on the rampage, shouting and yelling….great deal of firing, one of our Bren gunner over come by fear and was in his terror simply firing out'.[199]

'4th February 1944. Today all hell let loose, jap has attacked and infiltrated our lines, Jap planes have been over bombing and machine gunning,standing to all night.

5/6/7th February 1944. Jap planes and troops again, guns going all night, Indian troops behind us, got windy last night and fired into our camp for over an hour. No one hurt.'

From This position the Flight was unable to operate and was moved to the rear some fifteen miles at Bawli Bazar. C Flight under Captain Rex Boys was also in similar position in support of 7th Division

197 MaslenJones,Fire By Order,p-12.
198 MaslenJones,Fire By Order,p-13.
199 Ibid, pp,12-23.

at Mayu Hills, their landing ground was close to the 114 Brigade Headquarters located at the top of a hill. Boys undertook a sortie to gather information ,in the process he was attacked by the Japanese fighters, Boys suffered multiple fractures after crash 'he owes his life to the Burmese villagers who at great risk to themselves kept him away from the Japs'.[200] C Flight remained within the 'Box' for another three weeks fighting for its survival and rejoin the squadron when Ngakyedauk Pass was opened on 24th February 1944.British fought well in these isolated boxes and Air OP pilots regularly visited these for nothing but morale raising carrying documents on return flights.

After stabilizing of the battle the sorties for artillery were conducted, the pilots flying to the brigade headquarters,landing grounds being prepared by the regiments,pilot establishing frequencies and code words with the brigade majors of the brigades and carrying out shoots. 'it also happened very often that we would be asked by a unit to range their guns on to a feature.'[201] Captain Maslen Jones was for instance netted on four artillery regiments that include No.2 Field, No. 4 Field, No. 6 Medium and No.24th Indian Artillery Regiments.

The Journey of B Flight

Flight was initially left at Deolali for a proposed amphibious operation which never materialised.In February 1944 after the Japanese blow in the south came the main attack in the centre aiming at Imphal & Kohima .The flight was directed to join the 4 Corps at Imphal; it involve a journey of 1700 miles { London – Moscow}. Flight left Deolali on 16th February and with little knowledge about the country possessing only small scale maps they were able to cover the first 200 miles. This

200 MaslenJones,Fire By Order,p-18.
201 MaslenJones,Fire By Order,p-20

journey of the B Flight is a classic epic; the knowledge which they have imparted in terms of the native customs and character is an extension of the Alexander the Great's historians' record about India. Flight's journey is the first ever record of this nature in which they {pilots} unlike all their historical predecessors had the third dimensional view of the country, it is in fact a guideline for present day flyers as well. 'We had no idea from the small scale maps available as to what sort of country lay ahead of us and what were the chances of finding landing ground near the road. Having used known landing grounds to get us the first 200 miles and over the Western Ghats, it was another 1500 miles that we saw another recognized landing ground'. The road party would set off leisurely at about 9 in the morning which by Indian standard and keeping in view the urgency of situation was rather late and after a lapse of two hours the Auster would take off in pairs with half hour interval and only then the remaining ground party would leave.After passing the ground party the first pair would start looking for the landing ground close to the road about 30 miles on trying to find a place as far as possible away from any village, having found one the pair would land and the following pair would return back locate the ground party and after dropping the message regarding the location of the landing ground would also land and wait for the ground party to arrive which normally took another hour to reach'. However deserted a strip we choose by the time the last Auster had landed the Indians would appear in literally hundreds, bus trips were even run to come and see us as the news got around to the local villages by a remarkably rapid bush telegraph service. They would flock over to join us, firstly in ones and twos from the neighbouring fields,then on bullock carts, bicycles, tongas and decrepit lorries piled high with excited and bewildered locals. We would huddle the Austers into a corner and try to prevent their inquisitive fingers and sticks from doing major damage. At first we started by gestures and waving at them to keep away and not to touch. Their inquisitiveness always edged them in closer and closer, and our storming round shouting in basic Urdu and profane English merely encouraged them to feel and poke at one side while we were running round the other. Short of four pilots running round and round the Auster with big sticks for an hour, we would think of no other method of discouraging their over poweringly inquisitive fingers that played havoc with the fabric'.[202]

Finally the pilots found the political solution. 'If we could get the first arrivals to sit down cross-legged in a circle it made a cordon that held the others out, this we would by sitting down cross-legged ourselves and inviting the early birds to join us. Then a game of grandmother's step would start.The front row would begin to shuffle in, but surely. We would be offered bananas and various fruits as distractions, while the shuffling went on behind. Finally we would jump up and make a show of jumping on someone's feet, and amidst roars of laughter they would admit temporary defeat and back out a few yards. Providing we grinned and laughed and appeared friendly they could be controlled. To display temper was fatal, and threats with pistols and sten guns produced hollow laughter and egged them on…on one occasion we had difficulty in trying to keep a pair of elephants, well adorned with privileged spectators, out of trunk range'.[203]

Ultimately the ground party would arrive and this would give the pilots some respite but on the other hand it would create a commotion among the spectators as to the new acts of the strange drama being orchestrated in front of their eyes ,they would spent the entire night just gazing at the aircraft and discussing among themselves the theories of flight. At dusk the pilots would sit around their officers mess drinking chilled beer and smoking pipes and cigars where as the natives would

202 Unarmed into Battle,p-114.

203 Unarmed into Battle,p-115.

be puffing hookah.[204] The preparation of the meal the military drill was a magnetic attraction 'a cordon would be made by parking trucks in a circle and slinging the soldiers' and airmen's hammocks between. It was then possible for the sentry to patrol this fence and keep the crowds at bay...however damp or cold the night, by first light there was always a sprinkling of a crowd again...only in Bihar did we see signs of sullen hostility but no attempts were made to sabotage us.'The flight journey also paints the picture of the general state of military mind regarding the war. ' The country side varied enormously, and sugar fields or open parched plains provided reasonably easy strips. Polo fields though surrounded by trees, provided good pitches, and in the middle of Mhow when we picked on one we caused much consternation to the officers of an Indian cavalry Regiment still living in their little pre-war world. They looked with utter disgust at the ground party's dust-covered faces and road-stained jeeps, and complained that we were upsetting their evening 'chukkahs'. Once over the Central Indian plateau and into the flooded Ganges Valley 'We took to sending a jeep ahead with a shovel party to prepare the strips, as landing grounds became increasingly hard to find'.[205] The aircraft flew to their limits after taking from Benares, pilots were fascinated to see the funeral pyres and dead burning 'which they gazed down at as they flew up and down the city banks; a complete sacrilege to all good Hindus, as we learnt later'.[206] Flight reached Calcutta after a fortnight and took off for Imphal on 9th March.

The Japanese had launched their main offensive on **7th March 1944** to cross the frontier of India and seize the main supply base at Imphal thus destroying the British in the central front, it would have severed the railway line which was American General Stilwell's life line; it would have further allowed the Japanese to over run the

204 Unarmed into Battle,p-115.
205 Unarmed into Battle,p-116.
206 Unarmed into Battle,p-116.

airfields of Assam; thus cutting off the air supply to China. By the time B Flight was able to take off from Calcutta, Imphal & Kohima were both surrendered and besieged.

'Finally we left Calcutta...and flew up over 8000 feet mountains into Imphal Plains...it was real jungle country at last, with a wonderful sea of vast trees of every shade and colour. From the tops of the ridges one could see range after range spreading away across Burma towards China Hump. The sight on the way up of the Himalayas floating above the mist is one the Flight will not forget. At first all thought it must be cloud towering away to the north and it was with a sense of thrill that we realized it was indeed we had read and heard about'.[207] The flight spent the night at a make shift strip and next day landed at Kangliatombi a main strip five miles south of Imphal. Next day on 11th March 1944 early in the morning the flight was awoken by the Japanese Zeroes strafing the strip.

This provides an ample opportunity to put on test all the theories of AOP that have been raging for last four years. There was nothing wrong with the concept yet it was the rigidity in adherence to the tactical employment which came up here. The aircraft were unable to take off amidst the Japanese raids and shelling, thus for two weeks there was no operational sortie and finally the flight was pulled back to the corps headquarters which was fifteen miles back, the words of Air Marshall Arthur Barret proved prophetic again. The AOP historians and pilots have portrayed their deeds in this particular phase as exemplary because they were able to land all the nine aircraft including one Tiger Moth on a strip and evacuated nine sitting casualties, certainly an act worth praising but the hard fact is that similar American Air Ops were able to evacuate over two hundred casualties from the same area and that is where the British AOP is really put on microscopic analysis. The pilots

207 Unarmed into battle,p-116.

were brave but they lacked the appropriate aircraft and it was not RAF or Air Ministry fault because Army Council had them self been hasty in this.

American AOP's were employed by the Major General Wingate in his Chindit operations[208] for same deeds, British AOP did not took any part although their small aircraft would have been equally useful in those environments. AOP was practically prohibited from carrying out any other task other than the aerial observation thus casualty evacuation was never considered nor the aerial transportation of the senior commanders which almost all AOP pilots regarded as taxi service. Aerial photography, wire laying and air borne forward controller tasks were conducted later.

Captain Jones further highlight , 'It was always vital that I had an accurate understanding of the line of flight of the shell relative to my own position…I knew from information that I had been given exactly how long the shell would take to reach the target'.[209] Jones thus admits that a sound briefing is half the mission done. Captain Jones in his quest for intelligence briefing from the brigades had to fly at their position, an Engineers' effort would be expended to make a strip; Jones would then ride a horse for an upward journey as almost all brigade headquarters were located at higher ground. The Brigade Major would be having scant information themselves. In one of the very initial sortie which Jones conducted the chain of events went like this.

Hello Charlie 8, are you receiving me,over. {Jones}

Charlie 8 receiving you strength 5, over. {Baker Battery}

Charlie 8 Target registration map reference 418417. Fire by order,over.{Jones}

Charlie 8 Number 1 gun ready 12600, 32 seconds,over. {Battery, indicating that number one gun is firing at a distance of 12600 yards and

---

208 Personal experience in the 7th Division, Reginald Bailey

209 Maslen Jones,Fire by Order,p-29.

flight of shell is 32 seconds. The information that which gun is firing is important as the fall of round will depict the forthcoming scenario}

Fire, over. {Jones now fly on a line from where he can see the shell falling on the target,he has to behind the guns or away from the line of trajectory at a suitable height as well}

Charlie 8,12800, left 100 yards over.{Jones has observed that round has fallen short of target and too the right, he had to now bracket for distance and lateral line. A ground observation post would measure lateral corrections in degrees and minutes for translation to the gun sight, but an air observer does not have the luxury of time thus he gives his corrections in yards which meant that the troop commander had to make the calculation}.

Charlie 8 record as target-Mayu tunnels, west entrance over and out.

Japanese can monitor the transmission and thus in possession of potential danger, code words and nick names are thus use for this purpose and this is where the flight headquarters' role beomes important and along with it the liaison with the affiliated regiments.

Captain Jones was not employed in the all important artillery bombardment that took place on Razabil, 'I was not involved with fire for effect because the ground observation post had good observation, I was ask to have another look at the village and report on the effect of bombing'.[210] The most effective conducted by captain Jones was on a tiny village name Hintaya where Japanese infantry had taken position; it was a direct hit.

210 Maslen Jones,Fire by Order,p-31.

May 1944 Arakan Front

Weather started deteriorating with the onset of Monsoon and by end May 1944 the lone flight 'A' decided to call it a day and retreated first to Cox Bazar rest camp and later for a five month rest and recreation at Ranchi almost 250 miles north west of Calcutta almost 600 miles in the rear; ground party took three days to complete the withdrawal and Auster had to refuel at Comilla, Dacca & Calcutta before calling finals at Ranchi. The rest of the squadron also joined in at various dates. Captain Jones had to forced land close to Dacca because of Monsoon's threatening clouds, 'there is no way that an Auster could take on a tropical storm without being ripped apart'.[211]

Through out the Burma Campaign's, squadron headquarters remained with the 14th Army Headquarters in the rear, although in general playing no direct part in operations, it rapidly developed expertise in specialized functions, cable laying this indigenous method was invented by the commanding officer thus a pilot could lay out a signals wire from the Auster, this was a great help in the dense jungle terrain of Burma. At Ranchi the flight routine was leisurely. There were two cinemas in the town; Soldier Arthur Windscheffel watch ten different movies in fourteen days leave that was granted on rota. Another amusement at Ranchi was the ' Happy Valley', 'a spectacular act which was hosted by a lady of considerable proportions in costumes of minimal proportion depending upon the amount of encouragement from the crowd.[212] Corporal Denis Kemp wrote in his diary while on leave.

'September 2nd 1944, We are now climbing Mussori in a bus, roads very steep with sheer drop one side, driver seems a bit mad… having a wonderful leave, the air is very fresh like Blighty…saw the Himalayas covered in snow. Will be sorry when my leaves finishes. The

211 Maslen Jones, Fire by Order,p-36.
212 Maslen Jones,Fire by Order,p-42.

lady in charge is very good to us'.[213] Calcutta was another favorite and most sought after recreation town. Hotel Grand was the place to stay at, situated at Chowringhee, lounge was a great meeting place 'it was a large Victorian Palm Court complete with orchestra and dance floor, its restaurant Firpos was famous for dinning.

Imphal & Kohima; March 1944.

B Flight's two sections were in support of 20th Division at Kabaw Valley some 25 miles to the east of Imphal. These two sections were soon confined to the 'box' as on 21st March 1944 Tamu was abandoned and soon 20th Division also withdrew to Imphal along with the section. Imphal was also soon besieged and cut off, the fate of British India hang around this battle and it was the air power which saves the day by air insertion of 5th Indian Division which joined 17th, 20th & 23rd Divisions inside the Imphal. 'As the Monsoon broke all our strips near the divisions became flooded, and we were washed out. We then had to work from the two main all weather strips of Palel and Imphal, upto 300 Dakotas, Skymasters and Wellington were flying in daily bringing reinforcement…it was difficult to get a green from watch office for a mere Auster when half a dozen Dakotas were circling to come in'.[214] At times the pilots had to leave the shoot in the mid for refueling, this is something which is very unusual because the aircraft were operating from the main field and the battlefield itself was not more than twenty miles away. Pilots who came for refueling were seldom allow to take off again because of heavy air traffic. Captain Fowler had left the shoot in the mid for refueling ,first he was not allowed to get in the circuit. He had to fire a red Very light at the watch office to get in but was unable

213 Maslen Jones,Fire by Order,p-44.
214 Unarmed into Battle,p-117

to get away again 'Captain Fowler stormed into the watch office where a pilot officer told him it would have to wait until the empty Dakotas had gone. He was asked in no uncertain terms if he knew that seventy two guns were awiting for "that little thing" as he described it, it was suggested his empty Dakotas could wait. They did, and from then on the Austers had the priority over all but the spitfires on fighter patrol'.[215] The pilots also undertook propaganda leaflet dropping and mapping sorties. The battle was fluid in nature with areas rapidly being encircled by Japanese who often disguised themselves as refugees. Captain Southern once landed at a strip switch off the Auster and step out of cockpit when all of a sudden he realize that the position has been over run, he 'took off with his feet still hanging out of the door'.[216]

In April 1944 the squadron headquarters and C Flight left Arakan for Imphal and arrived at Kohima on 17th April.Relief of Kohima took place on 22nd June 1944 and even before that squadron had started preparing to proceed back to Ranchi for rest & refitting.

In this period a prolong paper battle erupted between the RAF elements and the AOP over the use of the aircraft, army channels fully back the AOP where as the higher command of SEALFA dealing with air matter supported the RAF point of view. The points ranged from having a medical officer for each flight to the installation of photographic camera on the Auster. RAF resisted it on the point that it is already being done by them and if installed on Auster will necessitate certain modifications matter was referred to Air Ministry which in turn again asked the RAF to forward comments which seems to be a height of military bureaucracy Squadron Leader Evans of RAF wrote that no firm opinion can be given on the suitability of F-24 camera in Auster IV aircraft without examination of air frame and the nearest aircraft are Auster III which are stored at

215 Unarmed into battle.pp-116-119.
216 Unarmed into battle,p-118.

Allahabad in crates, from the Auster publications the opinion of this section is, it is possible but there are doubts…installation might have to be done at the sacrifice of other equipment'.[217] The other points raised in the letter highlighted that since soon 656 Squadron will be replaced with Mark IV Auster thus we do not want to waste time on Mk 3 'we do not wish to waste technical experts time in examining the modification requirements when this can be obtained from UK'.[218] Air Liaison Officer{ALO} controlled the photographic section of the fighter reconnaissance squadron. 'we cannot create a special photo section for Air OP but we can provide them with K-20 cameras…my personal opinion is that there is little that the Air OP Squadrons can achieved photographically that could not be done as well or better by fighter reconnaissance'.[219] Finally Squadron Leader J.B Wales did allow only one aircraft for modifications to carry out the photographic missions after necessary modifications which again took considerable time.

The observer in the AOP became another point of tussle it entails extra weight which hampered the aircraft performance. AOP was again missing from the action once the decisive battle of Imphal and Kohima was being fought, it seems to be a pattern that at every critical junction the AOP was missing. The squadron was stationed at almost 1000 miles away from the area of deployment and its convoy had over forty vehicles and it took ten days of train and road movement to reach Kohima. There is absolutely no plausible reason why the squadron got itself stationed so far away, it was not the threat of the Japanese air attack, it is so unusual for a flying squadron to take this much time in reaching the operational area twice within a year.

'By the time we arrived Japanese were already pulling back'[220] Two flights were deployed {A and B Flights} in support of 33 Corps and C Flight at Arakan to support the 15th Corps squadron headquarters

217 Letter dated 17th June 1944,AIR 339.
218 Ibid.
219 Letter of Wing commander F.H.Issac RAF dated 4th December 1944 as part of Minute No.38
220 Jones, *Fire by Order,* pp.56-57

with HQ 14th Army. The A Flight was supporting the 11th East African Division and got itself established on a landing ground which was 11 miles away from the divisional headquarters, and later the flight made two more strips each involving considerable engineers efforts yet no decisive artillery shoots were taken., because of the ammunition supply the aerial artillery shoots were controlled by the BM of the brigade which means that the AOP did not had the executive control of the fire. The busiest day of the flight {22 November 1944} was when it conducted twenty three shoots with four aircraft on a narrow front of five miles using less than three regiments. After crossing of Chindwin the AOP was made to focus on narrow frontage each aircraft supporting one battery, which proved useful and this pattern remained till the end; fall of Rangoon. The other Flights also moved around involving considerable distances and time. Landing grounds kept on changing. AOP 'emphasis began to move from shooting to information sorties. Cable laying and photography were the principle services as well as some VIP passenger carrying'.[221] The squadron by end December 1944 was stretched to the point where one aircraft was supporting one infantry division. The other method of having concentrated effort at the corps headquarters was not employed where through communication the targets could have been sorted out by the corps staff. To fully understand the working of the AOP it is also relevant to have a look at the existing RAF pattern in the sector. The Army HQ and Corps HQ had the air liaison officers for the conduct of the air strikes but very rarely these AOP were incorporated into that pattern.

The issue of observer was raised by the 656 squadron in November 1944 pleading that 'observer will be there for look out for enemy aircraft'[222] and Air staff operation branch B2 informed them that as per policy RAF personal are not permitted to carry out these

221 Ibid p 86-87

222 Observer Minute No 34 dated 28th November 1944,AIR 23/5104 also see AIR 23/2552 also see WO 205/21.

operational duties and Air staff also seek the ruling from the medical branch on the issue. As per RAF procedure all personals have to be medically examined before they are allowed to fly as combatant air crew members[223] it is painful to note that at a time when the men were dying in the fields this unnecessary paper battle over a minor issue ensue because on 1st December 1944 the army inquired about the authority under which these airmen who have not been medically boarded and are not non commissioned officers can be employed on operational flying resultantly an amendment in the rules was made and 12 air men were allowed to be utilised for such purpose.

Winter 1944

It took eighty days to relieve the siege of Imphal. The two adversaries were for weeks on end in the closest possible contact 'literally the width of the tennis court which was beside the deputy commissioner's bungalow. On that tennis court many lost their lives including three brigadiers'.[224]

At the beginning of October orders arrived. The squadron was split into two, C Flight to move back to Arakan to support General Christison's 15th Corps supporting 25th Indian Division. A ,B & squadron headquarters to proceed back to Imphal to support General Stopford's 33rd Corps and squadron headquarters with 14th Army. Road party consisted of 40 assorted vehicles again covered 1000 miles of journey including 300 miles by train, crossing of Brahmaputra River. it took them 17 days. 'At the time we knew very little about what our troops had been through during the battles for Kohima and Imphal'.[225]

On 3rd November 1944 A Flight under Captain George Deacon

---

223 King's regulation 1446 clause 2{h} & 713 cover this aspect.
224 Maslen Jones,Fire by Order,p-55.
225 Maslen JOnes, Fire by Order,p-54.

including Captain Maslen Jones, Captain Pip Harrison and Captain Ian Walton and Captain Frank McMarth moved in to support 11th east African Division in Kabaw Valley; Kabaw literally means ‘death’. It harboured the most virulent form of any insect borne tropical malaise that could be imagined. As a precaution there was a daily parade of Mepacrin Drug for protection against malaria. The sections dug themselves in the jungles, the strip had to be made by the Royal Engineers using dozers and the ground defence responsibility by the infantry or the artillery.

Sections operated with two main task, to get the information about the fluid battle situation for the update of operational awareness of the staff, to locate the enemy strong points and direct fire power for own advancing infantry. Message dropping was another important feature, the messages were enclosed in in bags gay with red and yellow streamers which were used to enable the message to be found once it had fallen among undergrowth. The 11th Division contracted Scrub Typhus during its advance, apart from Malaria the other physical danger was the heat exhaustion for which the salt tablets were issued.

By mid December 1944 campaign was reaching its peak. Pilots were flying 16 hours in a day and at times taking over fifty shoots in a day. Squadron was supporting a front of 320 miles. On 19th December 1944 A Flight crossed Chindwin. Cable laying and photography were the principal services. On Christmas Day the Auster had a banner of twenty feet long painted in red with Merry Christmas flying over the own infantry positions and showering cigarettes.[226] The speed of advance can well be gauged from the fact that B Flight alone covered 700 miles of journey in December 1944.

The British counter attack of late 1944/45 was another opportunity to test the AOP, now the situation was more favourable for

226 Unarmed in Battle,p-125.

the employment of them as the air superiority was in the hands of allies. On 11th January 1945 A Flight was supporting the whole corps with one section each for two infantry division. AOP was employed in much wider and broader scope than merely artillery shoots and that proved useful, for instance on 20th January 1945 Captain Maslen–Jones was task to carry out the mapping of the 20 Division area of responsibility and on 7th February the two aircraft were in air for seven hours giving running commentary. One important feature of marked improvement in AOP was due to better communication between air and ground. In this phase the employment of AOP from LST was considered and War Office did send a copy and report of such venture by the Americans in Europe.

It was in first week of January 1945 that the AOP were used for the facilitation of air strikes where they indicated targets by having the smoke rounds fired on the target area. The AOP became important because it was carrying out a number of tasks, cable laying, drooping medical supplies etc. and it was appreciated by the army commander. The most notable feature was the first ever land fall from sea by the C Flight of the squadron from HMS Khedive. On 15th January 1945 HQ Air Command South East Asia wrote to Air Ministry requesting for two additional Air OP Squadrons 'existing squadron is now supporting 14th Army and 15th Corps'.[227] It was through out the campaign, quite amazing how all the squadron cooks consistently produced meals that had variety, ingenuity and about which there were never any complains.

On 12th May 1945 the whole squadron was at Rangoon. The Flights and the sections had varying roles in the advance of 14th Army they reached a crescendo in terms of flying hours. The squadron was more useful in the over all role of liaison rather than mere artillery observation.

---

227 Ibid

Table No.2Flying sorties of A Flight of 656 Squadron in Burma.

1944	Artillery	Reconnais-sance	Miscel-laneous	Hours Flown	Flight Sorties	Squadron Sorties
November	126	72	11	222	209	680
December	171	65	Not given	266	236	946
1945						
January	146	182	72	418	400	1251
February	376	98	10	522	484	1260
March	304	118	8	328	430	1238
April	166	85	25	235	276	847
May	52	38	4	104	94	282
Total of the squadron in-cluding three flights	2870	1484	2358			6712 sorties and 5710 hours.

Note. One sortie is counted whenever an aircraft takes off ,may it be for five minutes or an hour.It is obvious that 656 squadron maximum sorties were flown not in the support of artillery or reconnaissance but they are miscellaneous, the term by itself is unexplainable.

Chapter Six

Organisation & Administration

Air OP Squadrons were RAF units which were placed under Army Co-operations Command in United Kingdom, operationally under control of Home Forces. War Office used to allocate the squadrons to Home Forces or overseas theatres. Within the Home Forces these squadrons were attached with the corps by the MGRA and within the corps it was CCRA and in the division it was BRA who was controlling them. Its role was to act as an elevated air observation post with task to obtain information and to observe fire for the artillery. Its strength was equally provided by the RAF and the army{artillery}. In Indian theatre its strength was increased to cater for the bearers. Basic unit was the section designed to operate independently even cooking its own food.[228] One AOP Squadron was allotted to one corps and one to each army, in field they were operationally under the army control.

The basic organisation of the AOP virtually remained unchanged from beginning till end.[229] with some variation in man power in Indian theatre,[230] one squadron head quarters, three flights each flight having four sections and a flight head quarters. Squadron was commanded by the major and flights by the captains and so were the sections. Each section had one aircraft and four personnel's including the pilot, thus it was the basic unit which was mobile to some extent as aircraft could be transported in the integral jeep. Yet it was dependent upon the affiliated battery for administration and ground security apart from maintenance of strip.

228 WO Notes on the Air O.P squadrons
229 Artillery pamphlet no.7 deployment of an air op squadron, AIR 339 air op squadrons letter dated 22nd july 1944. see also; note book of AOP students Firepower museum
230 AIR/339 minutes no 14 ,8,26,28,32, 34 and 38 of year 1944.

This organisation was orchestrated by the Army Co-Operation Command and it was accepted by the War Office and artillery, probably this was the issue on which gunners should have done more brain storming but when one keeps in mind that these gunners had also been organised on to two battery regimental system till Dunkirk which proved totally irrelevant to the three brigade infantry division thun one can comprehend their lack of far sightdeness in this aspect also. The AOP remained under the control of MGRA who could have either put out the squadron in support of a corps or keep it under his control, CCRA would deal with the squadron as 5 corps had it 657 squadron under its command and than CCRA would put the flights to CRA at divisional level. On the other hand these AOP squadrons were RAF units and under command of army cooperation command and in theatres were placed under the RAF Group. Manpower was RAF in terms of technical and army for administration. The adjutant remained an RAF officer. In simple words it was a Pandora box and official correspondence reveals the same where issue of discipline 'can an army officer punish a RAF soldier' or 'can an army soldier go to town on RAF leave pass' were the daily petty issues.

Within a squadron there were two distinct divisions, one 'blue job' and other 'brown job'; the RAF were referred as the Blue. 'Our warrant officer, flight sergeants and fitters and riggers of various RAF ranks…absolute experts at their own particular tasks and had never learned to do the jobs of other experts, they were used to the comforts of an RAF station…on the other hand we had 'brown jobs'; a sergeant major, four sergeants and a variety of other ranks who were there to do all the other jobs in the squadron'. Well in these other jobs the conversion of a truck into a luxury caravan to be used in operations for Brigadier A.G. Matthew the commander corps royal artillery, 8 Corps was included 'even he later admitted that our squadron had done a superb job-at a cost to him of only fifty shillings for material'.[231]

231 Andrew Lyell,p-53.

In another incident two pilots of 658 Squadron {Sandeman & Bogod} were killed in an air crash when they hit a tree beside the River Avon near Larkhill, resultantly the commanding officer as a precaution 'frowned upon practicing evasive tactics by his pilots'[232] in another case one pilot had shot himself in flames by firing vary light pistol in order to warn another pilot of approaching enemy fighters.

The activities of an air op squadron were so varied that it was very difficult for even its squadron commander to say precisely what was permissible and what was not permissible.

Army officers were not happy with the RAF rules which includes that propeller of an aircraft can only be swung by the Fitter engine with chocks placed in front of the wheels, 'every flight had to be properly authorized and to be solely for training purposes and not for the convenience or pleasure of the pilot.'[233] But what if the training purpose happened to coincide with the convenience and pleasure of pilot? It was very difficult to draw the line and so when drawing it, 'we naturally allowed ourselves to be influenced by the morale of the squadron'.[234] But what is this morale all about, I have not seen a single entry in any pilot's log book or a mention in memoirs where a soldier was taken by air to attend the funeral of his father thus it is all about officers or more precisely the pilots moving around on weekends. The eggs round was a familiar sortie in which eggs were purchased by the pilots landing in the farm houses and purchasing eggs for the officers mess.' Having lived in the in Dorset and Somerset I could call on so many people I wanted to see and at the same time give instruction to pupils' remarked one squadron commander.

Air Ministry also passed instructions that no aircraft of Air OP was ever to land in any field which had not first walked on ground 'this

232 Andrew Lyell,p-57, Unarmed into Battle,p-161.
233 Ibid, p-62.
234 Ibid

was an order which could not possibly be obeyed if the Air OP were to continue to be of any use to the army.'

An incident highlights the RAF and military discipline standards, one day Captain Evelyn Predergast {651 squadron} was immediately told to proceed to Northern Ireland and there was no other way to inform his girl friend who was working in a hospital to be informed about him not keeping his date for the evening other than a message drop. Evelyn wrote the message and dropped it at the hospital where few people were standing, it happened to be a General carrying out the formal inspection, he referred the matter and RAF insisted on having the officer being court martialled for flying offense. Evelyn was able to get away on the fact that the day the offense took place the Air OP was still under the army rules and within these rules there was no mention of flying related offenses. 'I personally cannot see what Evelyn did wrong, it was not his fault that he was unable to keep that evening's date, no one would have objected if he had delivered the message in a jeep, a jeep journey would have taught him nothing; where as a journey by aircraft gave him some practice in message dropping'.[235] These are the remarks of an Air OP Squadron commander.

From RAF point of view the status of these AOP squadrons was not very different from other A.C squadrons.[236] It was responsible for command, administration, organisation including selection of aircraft and flying training. By July 1942, 651 Squadron was allotted to 5 Corps and 652 to Eastern Command, 653 to SE Command and 654 to Northern command and simultaneously these squadrons were also under control of RAF wings for instance 651 was under control of No. 32 Wing and 654 squadron under No. 33 Wing.

---

235 Ibid, pp44-66.

236 Artillery pamphel no.7 'deployment of an air op squadron dated 22 july 1944 also see AIR 339

Flying and general training was the responsibility of RAF it was Artillery school and AC Command which look after the technical aspects and finally it was CCRA who was responsible for the training of the squadron. War office selected the pilots and certain other ranks.[237] The important aspect of annual confidential report was also parallel where the ACR were written by the artillery chain of command and by the RAF as well. That is where the pit was located, no matter how much an officer is motivated but it's the ACR that finally decides the issue and that is why almost all AOP squadron commanders tried to have their ACRs written by the artillery seniors. Most of the problems and limitations of AOP originated from this root.

The basic structure of the AOP Squadron was formulated by the army cooperation command on 1st September 1941. For next two years this organisation remained subject of various conferences. apparently there was nothing wrong with the organisation as RAF made it on the basis of what AOP was supposed to do, it was to act in close cooperation with the supporting artillery battery and this is what War Office and artillery all along have been asking and promoting. From this point onward there were two dimensions of AOP, one in the form of the training flight at Old Sarum which was running the flying courses for the newly inducted gunner officers and the other being the 651 Squadron which was put under operational control of the home forces and attached with 5 Corps.

There is a difference between army and RAF life style and working environments, RAF had not forgotten how Bazeley had left them high and dry and at their own in France. The status of these AOP pilots were different from the normal army pilots seconded to the RAF in its army cooperation command, these pilots remained under RAF law

237 WO 233/7 dated 24th april 1944

and wear its blue uniform ,but in case of AOP the army and particularly the artillery took a stance that they should wear the army uniform and moreover the squadron should be commanded by the artillery officer. This was something which had no operational bearing at that time.

Flying Tactics.[238] Basing upon the trials it was restricted to fly at maximum to 600 feet, not remaining airborne more than 20 minutes and not to fly within 2,000 yards of occupied enemy territory. Communication was mainly through the aircraft set No 21,which could provide short range two way communication with artillery battery, another set No.9 was for communication between flights and squadron headquarters. Its tactics and doctrine was responsibility of artillery school and A.C. command.

Responsibilities[239]

RAF. Administration, organisation, including selection and provision of aircraft, general and flying training

Army. Operational Control, selection and provision of pilots and certain other ranks personnel, certain equipment

Training

Training of squadron was responsibility of A.C.Command only in terms of flying and general where as School of Artillery was responsible for artillery training and in formations it was CCRA's responsibility.

---

238 Note set before army Command meeting at GHQ Home Forces on 13th august 1942 by A.O.C in C. Army Co-operation Command,Appendix 'A' to HF,11487/Ops dated 29.7.42

239 Ibid.

Out Line Organisation[240]

One squadron headquarters which had four reserve aircraft and two pilots commanded by a major, Three Flying Flying Flights A, B and C, each commanded by a captain and one servicing flight which had one mobile maintenance party and one static workshop commanded by a RAF officer. Each Flight had four sections{A Flight 1-4, B Flight 5-8 and C Flight 9-12}

Officers.	***RAF***	***Army***	***Total***
Squadron Headquarters	2	3	5
Servicing Flight			
Three Flights{ each]	-	15	15
Grand total	2	18	20

Other Ranks	***RAF***	***Army***	***Total***
Squadron Headquarters	18	22	40
Servicing Flight	23		23
Three Flights{ each]	14	16	30
Grand total	83	70	153

Aircraft. Initial equipment 12 Auster, Four reserve held with the HQ Flight.Thus in a flight there were five pilots including the flight commander but only four aircraft. Squadron commander also had no aircraft at his disposal other than to fly one of the reserves..

The first outline organisation was formed by the commander of 70 Group and it was based upon on correspondence with School of Artillery. It was structured on having one aerial observer for each artillery regiment, thus 14 Air OP's for a corps having three division. and one central servicing station of RAF.[241] Each Air OP to consist of following.

One Pilot officer of RA, Three other ranks{flight mechanic, flight

240 Ibid, appendix A. also refer to WO Notes on the Air O.P squadrons

241 AIR 39/47; Letter from No.70 Group,RAF to Army Co-operation Command,No.70G/S.751/Air dated 18th December 1940

rigger and driver}, One light aircraft, One lorry to carry fuel,lubricants and aircraft stores, One Motorcycle for inter-communication between command post and landing ground.

656 Squadron operated in Burma with minor changes which include having one medical orderly for each flight, one batman for each officer, mess waiters were also included in it.[242]

656 Squadron organization was as under.[243]
23 Army officers Royal Artillery all pilots.
3 officers from RAF all non flyers, adjutant from RAF.
90 Gunner soldiers Royal Artillery
80 Airmen,RAF.
Total ,196 all ranks.

Squadron Headquarters

Commanding officer, Major RA.
Second in Command, Captain.
5 Pilots all captains RA.
Adjutant, RAF.
Equipment officer, RAF
Administration officer, RAF.
Non Commissioned Officers & Other Ranks.
Army. 42{Motor transport, signalers, drivers, medical orderly}.
RAF, 38 {engine, airframe, instrument and wireless mechanics, cook}.
Three flights name A, B & C ,each having a headquarters and 4 sections.

Flight Headquarters.

Flight Commander, Captain RA. 8 other ranks from RA{ Military transport drivers, medical orderly, cook}. 6 other ranks from

---

242 MD artillery Training Volume 1-General Tactical Employment,Pamphlet No.10.Employment and Organisation of The Air O.P.p.2.sssssss

.243 MasleyJones,fire By Order,p-172.

RAF {engine & airframe mechanics and wireless operator}.

Communication-No. 22 set netted to local command headquarters, section pilots and squadron headquarters.

A section of 656 Squadron consist of one captain of Royal Artillery, one aircraft fitter airframe and one aircraft fitter engine both from RAF, two gunners one acting as signaller and one as driver/batman; both from Royal Artillery. One aircraft, one truck / jeep with trailer. Section had communication set No.22 netted to flight headquarters and also with the headquarters of artillery regiment to whom it supported.

AOP Pilots Training

This aspect is important as it throws the light on the controversy of RAF and army rift and the constant theme of RAF being opposed to AOP. It is by going through the detail aspects of flying training that one understands the delays in raising of squadrons.

Training of AOP was a dual responsibility between the RAF and the army.[244] AOP was exclusively an officers affair. RAF had its own unique selection and training procedure where candidate was groomed into flying culture from the beginning and as such the standards were very high, in one study only 25% candidates[245] would pass through the Elementary Flying Training {EFTS} before they were moved into higher training. Airmen have been regarded as members of an elite group, RAF selection and classification of air crew generally depends upon the natural aptitude for flying and also on the socio-economic back ground.[246]

244 AIR 39/70 flying training of AOP pilots policy,also AIR 39/32 conference AC command 18th April 1941 at old Sarum.

245 Well.Mark, & Courage *Air warfare the allied aircrew experiences in the second world war*{Frank Cass,London,1995} pp 13 also see pp 4-12

246 Ibid

AOP by its design was based upon the assumption that the officers would all be proficient in gunnery affairs and as such all they required is basic flying training. However both these assumptions proved wrong. The officers all volunteers were selected by the War Office but in the first two courses and also later on the recommendation of Major Bazeley played an important role and the very first officer he trained was a lieutenant who in turn was given a carte blanche by him to select the first course and he in order not to loose his seniority selected majority of lieutenants and few captains out of which one was dropped because of this factor, it was only in later years that selection procedure was streamlined. Army kept RAF at bay in selection of pilots and thus those officers were also selected who had no aptitude for flying and later many of these were dropped and time on training was wasted.

RAF adhered to their flying standards and rejected the concept of accepting the civil flying club qualifications or standards. Under RAF the officers underwent the EFTS of twelve weeks and later another twelve weeks at 43 Officer Training Unit{OTU}[247] which was solely established for this purpose at Larkhill. Initially the D Flight [1143 Flight] was responsible for this and later it was upgraded into 43 OTU. The instructors were AOP and RAF pilots. AOP pilots were sent to flying instructors school for instructor rating as well. The ground subjects was the responsibility of Artillery School at Larkhill, the last part of the training was solely for the artillery shoots and the students were trained in the art of artillery shoots on miniature range as well. Thus the students had to under two different types of training under two different types of organisations. The gunnery syllabus at No. 43 OTU comprised of four weeks for those officers who were not qualified as gunners, there were three lectures by the instructor gunnery{IG}, 10 minutes range periods daily and two lectures by chief gunnery instructor on Air OP gunnery

247 Log book of Captain J.H.Creswell ex 17 EFTS,Fire Power Museum.

procedures with maximum of four air shoots.

Flying training's final evaluation was the domain of RAF and surprisingly it was the AOP squadron commanders who later wrote to RAF requesting them not to give good grades to the pilots as it creates sense of over confidence in them and with statistics they pointed out that majority of pilots having such grades have met accidents in initial days of squadron flying.[248] From the course note books of students it is obvious that they were not very proficient in artillery or general knowledge about the army working, as majority of them were lieutenants. They were taught tactics and organisation of division as well. The most glaring aspect of artillery training imparted was reliance upon the 'clock ray' system of indicating targets the very basis on which AOP was raised although the standard procedures were taught yet it was emphasised that 'immediate neutralisation of target is important'.[249] The role of artillery school needs scrutiny as being non flyers they insisted upon adherence to the ground procedures the 'Gospels of Larkhill'[250] were difficult to change and AOP were stressed 'don't try to change the existing procedures' thus they took away the initiative from the pilot in the air and made him bound to the rules.

There was no training for the observer or ground troops for peculiar AOP operations in terms of strip preparation or protection and it was all imparted in the field on daily basis. In European Theatre the quality of training was reflected again and again ,it was a standard procedure that any officer coming from the Larkhill still had to undergo the operational training on the spot before he was cleared for operational flying.

Initially Major Bazely and Lieutenant Colonel Hilton toured the artillery formations giving lectures and motivating young officers

248 Letter from HQ No.32 Wing RAF to HQ Army Co-operation Command dated 3rd April 1943,AIR 39/70 Flying training of AOP pilots policy.

249 Ibid

250 Ibid

to volunteer for the Air Op.[251] In early November 1940 No. 9 EFTS at Woodley near Reading accepted 35 army officers with four or five gunner officers for AOP training the gunner officers were not permitted to be seconded to the army co-operation squadrons, it should be kept in mind that RAF had the army officers being trained in flying and posted to army co-operation role, these officers came from varying arms and regiments and it were they who were taking the artillery shoots but Royal Artillery and army insisted on having thoroughbred gunners flying the light un armed aircraft exclusively for artillery shoots but later events in all theatres of war exposed this fallacy because Air OP were utilised for casualty evacuation, aerial photography, transportation of commanders and artillery immediate neutralisation all these tasks were already being performed by the army pilots being seconded to RAF thus duplication of efforts. For the training of flying instructors in Air OP a two month course was run at Cambridge at Flying Instructor School. Pilots were given two shillings per day as flying pay. The issue of flying instructors in AOP was a delicate matter,the problem of instructor being junior to the students in military rank was a serious issue in which at least one student Captain Hugh was returned to his unit from flying for the reason that instructor Lieutenant Ralph was junior to him 'Ralp gave us captains a rough time…Captain Hugh once said that was not going to be spoken like that by a subaltern so Hugh was failed'.[252]

Elementary Flying Training School

80 hours, normal circuit, low level navigation, forced landings.

Officer Training Unit

10-12 weeks

---

251 *Memoir of Air OP Pilot*,p-11.
252 Ibid,p-25.

Cross wind landings.3 hours
Instrument Flying. 2 hours
Night Flying. 3 hours

In February 1943 Form 5011 was introduced for Air OP pilots in which flying tests were classified as general flying having maximum of 800 marks out of which low flying had 200 marks precision landing having 200 forced landing having 100, airmanship and navigation had 100, turns having 100 and aerobatics also having 100 marks, so did instrument flying with 100 marks, night and link trainer flying also had 50 marks each.[253]

The army pilots mind set is obvious from Major Andrew Lyell's remarks 'we could have become expert fighter pilots far more easily than an RAF trained pilot could learn our job'.[254] The resentment of army pilots range from the RAF instructors not being able to carry out short landings to the fact that RAF instructors would take their pets into air 'when his pet whippet required exercise, he {Squadron Leader Donald Walker} would take it flying with him,land somewhere on Salisbury Plains and give it a chase after hares'.[255]

Another bone of contention was the flying techniques being followed by the RAF, army pilots would recognize the senior RAF officers by virtue of their landing patterns 'a group captain was expected to do one bump landing, an air marshal a two bump landing and an air chief marshal a three bump landing...Air Chief Marshal 'ugly'[256] Barrett was said to 'stir the pot' during glide before touching down, a practice adopted by pilots during the WW1 of moving joy stick round and round in a circle in an endeavor to shorten the length of the float'.

---

253 AIR 39/70 Flying Training of AOP Pilots.
254 Andrew Lyell,Memoirs of an air Observation Officer,p-49.
255 Andrew Lyell,p-49
256 Ibid,p-51.

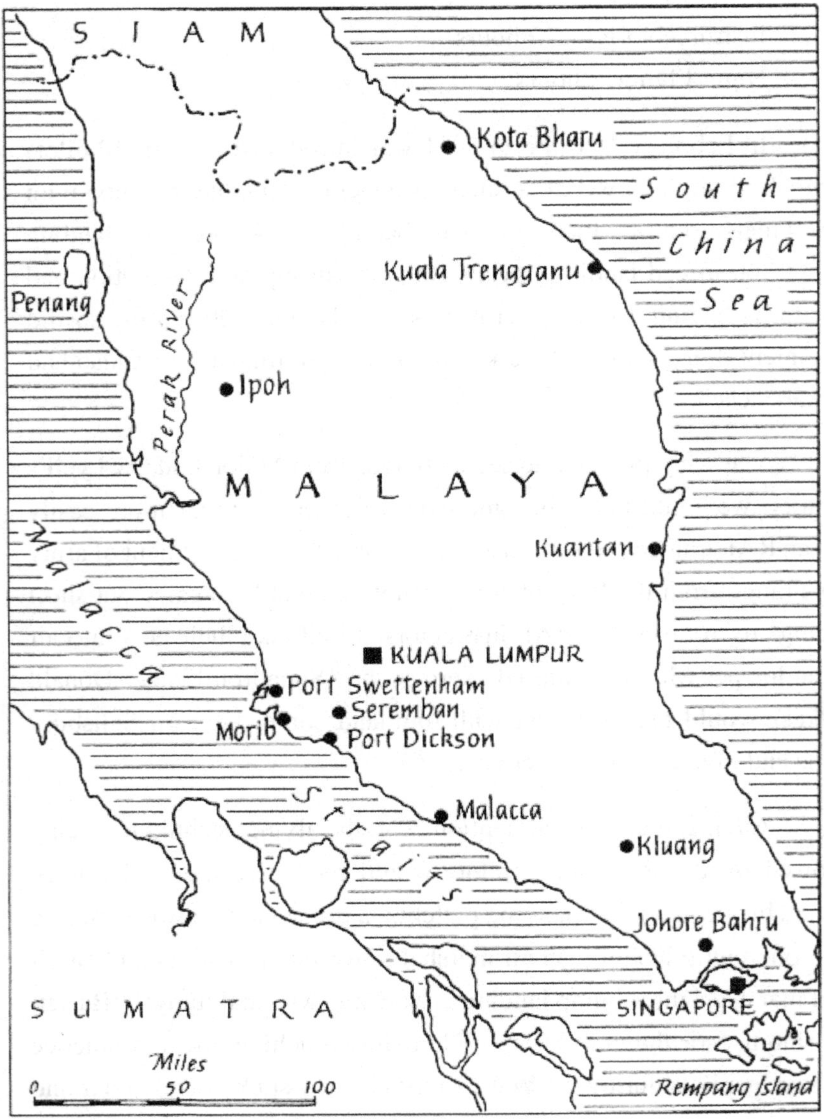

Map Courtesy **"FIRE BY ORDER".** By **E.W.MASLEN-JONES**

Chapter Seven

Post World War 2

American bombing took lethal shape with the development of B-29 bomber, and in June 1944 USA was able to attack the Japanese mainland from China, but it was again based upon area bombing rather than precision bombing. It was only after capture of Marianas Island that USA had a secure base to launch attacks .There focus was again on civilian killing, on 24th February 1945 over 175 B-29 bombers bombed Tokyo with incendiary bombs destroying one square mile area and killing over fifty thousand humans. The war ended on different notes at Europe and Asia. in occupied Germany the four western powers had their own areas of spoil The American, British, French and Russian. The war in Europe officially came to an end at midnight on May 8th 1945, in the final collapse of Germany, Field Marshall Montgomery led the main striking force. Germany was divided into four occupation zones with Britain occupying the northern Germany, respectively the field commanders were designated as the military governors, Field Marshall Monty was the first such British governor, later from 1950 onwards these were designated as high commissioners.

Germany alone suffered 2,050,000 civilian casualties along with a further 2,010,000 people wounded or permanently disabled, Japan suffered 393,400 killed and 275,000 permanently wounded or missing civilian population, United Kingdom suffered 148,000 including 60,000 killed USSR had 6,700,000 civilian casualties and USA had nil. The percentage of UK soldiers having battle wounds from aerial bombs is

75% as compared to 10% from land mines and booby traps.

Can this war be won by bombing alone or through air power? This was a common question in Britain during the war itself and none other than Air Marshall Trenchard states 'I do not know, I have never claimed that we can. Equally I have never suggested that we cannot'.[257] In this phrase lies the whole history of air power and its impact on the outcome of second world war.

In 1943 a total of 200,000 tons of bombs were dropped on Germany nearly five times as much as in 1942, yet German armament production overall rose by 50% in 1943. The German fighter aircraft production after a drop of 25% in August 1943 again rose to new heights in 1944. The German generals who were captured after the war do admit like Field Marshall Kesserling, 'that Allied strategic and terror bombing behind German lines along with tactical interdiction were the three major causes of German defeat in WW2'. Field Marshall Rundstedt stated 'I believed that the greatest factor in Germany's defeat was the strategic bombing by the allied air force particularly during Ardennes offensive…destruction of railway'. Field Marshall Model in his 1st November 1944 orders admit 'Hostile air force has made movement to the rear impossible'.

By mid 1944 out of 487 squadrons under RAF command 100 were provided by Royal Canadian, Royal Australian, South African, Indian and Royal New Zealand Air Forces, out of total air crew of 3,40,000 the 1,34,000 was provided by the empire.[258]

After VE the British armed forces set to occupy and reconquest the empire lost to Japan in Asia and in some cases to help fellow European nations in getting their colonies back, yet a wave of nationalism had

---

257 *Four Essays of Trenchard,Royal Norweigian Air Force Journal Volume 17*

258 Right of Line,p-351.

started in the former colonies which after their armed resistance to the Japan were not very conducive in welcoming back their European masters, thus the initial period immediately after the VE is critical for the future of British empire as a series of chain of events involving military. political and economical factors shaped the future.

French Indo China

20th Indian Division under General Gracey intervened in Indo China in September 1945 and on 30th November 1945 Admiral Mountbatten accepted sword of surrender from Japanese at Saigon and by mid January 1946 the 20th Division started pulling out.

Indonesia, Dutch East Indies

First major British landing was in Jakarta on 15th September 1945 the 23rd Indian Division.[259] Surabaya, the largest naval base in Asia after Singapore, here 29th Brigade of 23rd Division arrived on 25th October 1945 and by 29th October British had lost 16 officers and 217 other ranks while fighting in the city against the civilians, White Hall announced the British garrison is besieged[260] British commander Brigadier Mallaby was also killed in an ambush and resultantly a reign of vengeance spread out, 5th Indian division was also called in for help, the city refused to surrender and local Muslim population raised jihad, Surabhaya became the first and largest British military engagement after WW2,city was bombed by RAF. On 28th November 1945 the city was cleared with a loss of 600 allied troops and some 10,000 civilian

259 C.Bayly, and T. Harper, *Forgotten Wars,The End of Britaiin's Asian Empire*{London;Penguin,2007} P,167 ,
260 Ibid., 177.

casualties. Feeling of hate spread all over Indonesia, in one instance a crashed British Dakota with 23 British and Indian crew were hacked to death and in retaliation British burnt over 600 houses. British left Dutch East Indies in November 1946.

British had 2,136 own and Indian troops casualties in war against nationalist in Indo China and Dutch East Indies

Operation Zipper. Malay peninsula is roughly the size of England less Wales, located in South East Asia, it is a jungle mountainous area, the maximum elevation of mountains is 7000 feet and they bisects the country into north-south direction, jungles covers four fifth of Malay's 50,850 [261] square miles of area, the population of Malay in 1953 census was 5.2 million [262] people in which Indians constituted 11%,Chinese 39% and rest were the native Malay race, these Chinese and Indians mainly came to Malay because of rubber and tin trade that flourished since British took colonial control of Malay in 1895.[263] In 1925 the communist party of Malay was formed and it was this very party that became the sole resistance against Japanese invading army that over ran Malay and Singapore in 1943. These communist guerrillas were contacted by Major John Davis of British Army in 1943[264] and weapons, ammunition were para dropped into jungle, few British officers and men also joined these communists to give them training in weapon and explosives, this force came to be known as 'Force 136'.[265]

Squadron 656 commanded by Major Dennis Coyle was stationed at Madras at end June 1945 from where they moved to Attur, the officer commanding forward a report at the end of month in which

---

261 Noel Barber, *War of Running Dogs, How Malay Defeated The Communist Guerrillas 1948-1960* {London, Fontana 1972}p 11

262 Geoffrey Fairbairn, *Revoloutionary Guerrilla Warfare* {Suffolk,Penguin,1974}p 160

263 Robert Jackson, *The Malayan Emergency and Indonesian Confrontation, The Commonwealth's Wars 1948-19666* {Yorkshire,Pen and Sword,2008} p 5

264 Margaret Shennan, *Our Man in Malay* {Glouchestershire,Media Group,2007} p 60

265 Ibid p 75

he highlighted the unsuitability of the location,wisely Dennis placed himself under command of Royal Naval Air Service at Coimbatore,it was here that with in three weeks {end July 1945} all the six Auster of squadron were over hauled, main planes were re-covered with linen fabric, remaining 10 Auster were at Madras which were also repaired and serviced with the help of 139 repair and service unit. One detachment of the squadron sailed in landing craft {LCT} towards Malaya where an opposed landing was anticipated, another detachment of the squadron was at Bombay waiting for convoy which would leave for Malaya in last week of August, seven pilots were assemble at Trincomalee {Ceylon} to sail with the first wave on HMS Trumpeter a Royal New Zealand ship. 656 Squadron was allotted to 34 Indian Corps, which had placed A Flight in support of 23 Division,B Flight {two sections} with 5 Division, C Flight with 25 Division.

Major Dennis raised the point of exhaustion and fatigue of the squadron highlighting the fact that Air OP Squadrons were available in European theatre yet they are not being utilized in this theatre 'squadron has undertaken the work of three{squadrons} for the last two years with the result that there has been bare minimum of rest'.[266] The Austers were lifted into the Trumpeter and on 21st September 1945 they sailed towards Malaya.

On 24th September 1945 at 1000 hours the two Austers took off from the Trumpeter, pilots were given a compass bearing and they had to fly almost 45 minutes over sea, six men were assigned to hold the Auster from wing struts and tail plane until the last moment on deck as Austers were extremely vulnerable to strong winds at landing deck, handlers placed the aircraft in the take off position and held it till the last moment when pilot gave the signal for release, Captain Maslen Jones was the first one to take off with Captain Mike Gregg, Jones had his

266 Fire By order,p-150

complete kit in the back of the aircraft,this has been a regular feature of army pilots where they carry their heavy personal kit alongwith which increased the take off roll 'I decided to use the height of the deck above sea level to gain additional air speed,after which I climbed to 1000 feet and set course for the main land'.[267] Austers landed at Kuala Lumpur golf course where the head quarters of the squadron was established and all Austers were parked, a skelton ground crew were also flown in by the commanding officer and his second in command, the remaining part of the squadron joined in after three weeks. C Flight moved to Ipoh area, A Flight in Seremban and B Flight at Kula Lumpur. The main fear of the squadron was that they would be used for air transportation of senior officers which was controlled by the HQRA 34 Corps.[268] Squadron conducted an aerial photography of the Singapore island which took them a week. One quite interesting and historical sortie was flown by Captain Maslen Jones with General Seishiro Itagaki 7th Japanese Army commander who had surrendered three quarter of a million Japanese soldiers to Admiral Loius Mountbatten on 12th September 1945,there were 300,000 Japanese men on the island and it was decided to put all of them on a near by Rempang Island, he was allowed to reconnoiter the island to assist in his plans for the movements of his troops. On 25th November 1945 the sortie was flown with only the pilot armed with a revolver and the surrendered general in the Auster.

The post war period in Malay became very politically agitated over future of Malay, communists who were dominating the trade unions almost paralysed the economy through strikes, the British were mainly the owners of large rubber plantation farms that were scattered all over the country, Chinese were the workers and squatters of these farms, Indians were into business.[269] On 18th June 1948 emergency was imposed

267 Ibid,p-154

268 Ibid,p-156

269 Jhon Nagl, *Learning To Eat Soup With a Knife Counterinsurgency Lessons from Malay and Vietnam* {London,Chicago University Press,2002} p 63

all over Malay and it officially lasted till 30th July 1960,[270] although in between Malay it self got independence on 31st August 1957.

This emergency, insurgency and guerrilla warfare is a matter of interest from various angles, the first and foremost was that it was not allowed to spread the way Vietnam and Algeria got out of hand, and is regarded as the only victory of west over communism thus British handling of this emergency against communist became a kind of standard yard stick on which all other are judged especially the way French and American got bogged down in Indo-China in same time period and almost same is happening now in Iraq and Afghanistan.

Total security forces losses were 1865 killed and 2560 wounded where as civilians losses were 2473 killed and 810 wounded, the rebels were 6698 killed, 2819 wounded and 2675 captured.

Deteriorating Morale, Mutinies

The major worry for British force commander Admiral Louis Mountbatten was the deteriorating state of morale, he told Chiefs of Staff in October 1945 'morale is good but it would deteriorate'[271] In October 1945 there was a mutiny aboard HMS Northway where sailors left their breakfast ,protesting over the menu, in January 1946 there were series of protest across RAF stations starting from Karachi, it involved 14 stations and over 50,000 men,[272] immediate cause was poor food and living conditions and by May 1946 it spread to frontline troops in Malay as well where in one instance over 258 British troops were arrested and charge with mutiny and 243 were sentenced, to the military it was the bombshell. Montgomery wrote to field commanders 'no criticism

270 Ibid p 103
271 Ibid, 171.
272 Ibid ,218.

against our new government'. In February 1946 RN had mutiny over racial tension and soon anti British riots erupted in India.

None has influence British way of war in post WW2 as much as the first Labour premier, Mr Atlee's policy and thinking was to avoid any clash with Russian and to withdraw forces from all such areas including Middle East and India and to have a defence line across the Africa. He was the first to grasp the fragility of empire in post war era 'Its not a unit that can be defended by itself, it was the creation of sea power and with the advent of air power…empire scattered over five continents cannot be defended with a fleet based on island fortress, its gone'[273]. Mr Atlee instead came up with the idea of having a commonwealth of Britain former colonies. Poor financial conditions of economy and social programme of his could not afford traditional strategic culture of empire and confrontation. By 1946, UK had to take a 3.75 billion $ loan from USA, yet UK was the highest spender on defence surpassing even USA.

It was in these backgrounds of ideology and economics that Labour announced the forthcoming independence to India in 1947 and later to Burma as well. Military did not agreed with the prime minister and in January 1947 the three chiefs stressed the retention of Middle East bases and influence,

Jewel in the crown was India which was also strategically vital for any counter threat to communism. In India the talks amidst increasing street violence and ethnic killings finally resulted in India being divided on religious grounds with pre dominant Muslims having East and West Pakistan on the eastern and western frontiers of Raj India, both countries were dominion in nature with India becoming a republic in 1948 and Pakistan in 1956.

---

273 Olav Riste, ed , *Western Security the Formulative Years 1947-1953*{New York; Coloumbia University Press,1985},p.32.

Flying came quickly in Sub Continent, the world's first official airmail was carried on 18 February 1911 when French pilot Henri Piquet flew Humber biplane from Allahabad to Naiai Junction across River Jumna a distance of 5 mile with 6500 letters. The First Indian Ace, **Indra Lal Roy** was the most accomplished of all Indian pilots who flew during the First World War. Born in Calcutta on 2nd December 1898, this gallant flyer was only 16 when the war broke out in Europe. He had been a student at St. Paul's School, Kensington, London since May 1911 and left to sign up for the Royal Flying Corps as soon as he reached military age in April 1917. He was granted a commission in the service as 2nd Lieutenant. (General List) on 5th July 1917, a few months short of his 19th birthday. Five days after commissioning into the service, Roy moved to Vendrome for flying training. And after a stint at the Gunnery School at Turnberry, Roy was posted to No.56 Squadron on 30 October 1917. Between the 6th of July and the 19th of July 1918 the 19 year old was credited with ten victories. In just over 170 hours of flight time. On the morning of 22 July 1918, three days after scoring his final victory, this gallant young man was killed in action when his plane went down in flames over Carvin during a dogfight. He was posthumously awarded the Distinguished Flying Cross on 21st September 1918, the first Indian ever so honoured two other Indians were also given commission in RFC they were HS Malik {the very first Muslim flyer of India} and Lieutenant Welinker. In 1914 a Central Flying School was established at Sitapur it was later abandoned and three above named officers who were under training in England joined RFC detachment at Mesopotamia in November 1914, the remainder of the School was sent to Egyp.In 1915 RFC's No 1 Reserve Squadron formed the Nucleus of 31 Squadron which arrived at Bombay on 26th December 1915 with five BE.2C, it's A flight flew to Nowshera and later moved to Risalpur on 1st March 1916, in November 1916 the squadron moved to Risalpur with flights operating at various places.

By the end of Great War, Royal Air Force had assume the policy of deterrent through airpower in Middle East and North Western Frontier thus further squadrons arrived in India equipped with DH-9A air craft by mid thirties there were six such squadrons. In 1918 first flight from England to India via Egypt took off on 29 November and landed at Karachi on 12 December 1918, crew consisted of Captain Ross and Major General Salrond. Peshawar, Kohat, Risalpur and Karachi were the initial air fields. Razmak{1935}, Miranshah, Samungli {1922} were all constructed before the Second Great War mainly to counter the tribes as well as being part of Great Game. In 1921 a flying school was opened for a short time at Quetta equipped with Bristol Fighters and Avro 504Ks. Also on 28th March 1921 the Indian legislative passed the bill for the gradual Indianisation of armed forces.

In 1929 the world's first mass evacuation of civilians took place when 549 people were evacuated from Kabul to Risalpur. Regular commercial flight from England–India was inaugurated on 30 March 1929; it used to took seven days for 4130 miles. In 1930, the year Lahore Flying Club was Opened, six Indian officers were sent to England for flying course, the first five pilots commissioned into the IAF were Harish Chandra Sircar, Subroto Mukerjee, Bhupendra Singh, Aizad Baksh Awan and Amarjeet Singh. A sixth officer, S N Tandon had to revert to ground duties as he was too short. All of them were commissioned as Pilot Officers in 1932 from RAF Cranwell. The Indian Air Force (IAF) was established with the passing of the Indian Air Force act on October 8, 1932. Accordingly, IAF's No.1 Squadron came into being on April 1, 1933[274] It was formed as 'A' flight of IAF at Drigh Road Karachi; the same year Mount Everest was first flown over by Flight Lieutenant Intyre and Marquis of Clydesdale. Initially, the IAF consisted of five

---

274 Ray Sturtivant, *Aviation News,24 February -8th March 1984*,pp-884-889.

Indian pilots, one RAF Commanding officer, Flight Lieutenant (later Air Vice Marshal) Cecil Bouchier and four Westland Wapiti biplanes serial number J9735, K9854, K1263 and K1272. At that time there was a prerequisite of 39 inches leg length as medical standard for the pilots. Simultaneously 22 technicians were also selected as having railway work shop experience as enough to handle aircraft, they underwent technical course at Karachi on 19 January 1932. These technicians were known commonly as "Hawai Sepoy". In April 1936 the B Flight was formed and A Flight was moved to Peshawar where already 20 Squadron RAF was deployed. The No 1 Squadron later moved to Chaklala in November 1936 to join No 5 Squadron RAF, its flights operated at Miranshah finally the squadron assembled at Ambala on 3rd July 1938, its C Flight was raised at Drigh Road Karachi in the previous month.

In September 1939 the second Great War began and expansion of Indian Air Force took place, a flying training set up was established at Risalpur. The training pattern of IAF also included an initial training of 10 weeks for the cadets at Lahore {No 1 Initial Training Wing} where they were taught drill and military discipline later it was transferred to Poona and duration was increased to 14 weeks and extended to 18 weeks in 1943. Abinitio flying training was the responsibility of No. 1 & 2 Indian Elementry Flying Training Schools located at Begumpet {Hyderabad} and Jodhpur respectively, the duration was 10-12 weeks in which ground subjects and flying training was imparted. From here the cadets would move to No. 1 Indian Flying School at Ambala where cadets initially underwent ten weeks of flying followed by twelve weeks of advanced flying training and then posted to the squadrons, civilian flying schools were also utilised for training of IAF pilots, another small initial flying set up was established at Risalpur with few Tiger Moths but it was disbanded in 1941. An armament training unit was formed at Drigh Road for providing training for pilots and observers but this was also closed in 1941 and training was shifted to Peshawar where a new No. 1 Armament Training Unit was established, a parachute training school was also established on 22nd September 1941 near Delhi at Willingdon Airport initially known as Air Landing School. Japanese attack on Burma and air raid over Calcutta forced the British to built over 200 new airfields especially in depth of India thus Dhamial, Rahwali, Gurkha, Multan, Fatehjang, Sargodha, were constructed in this back ground. In 1942 all civilian flying schools were closed and aircraft taken over for anti aircraft cooperation flying[275] on 1st April 1942 No 1 Operational Training School was set up at Risalpur for conversion of Hart and Lysander aircraft pilots onto Hurricanes and in November 1942 No 152 OTU was formed at Peshawar to undertake bomber conversion, training commenced in January 1943 and by May same year the unit had 33 Venegeance and six Harvard aircraft along with six target tugs,to cater for heavy bomber training No 1584 Conversion Flight was established at Salbini near Calcutta

In India one training flight for AOP was established at Deolali it was known as **1587 Air Observation Pilot Refresher Flight** whose postal address was c/o RAFPOST, Bombay. It started operating in November 1944[276] and in two years over thirty officers including one Indian artillery officer 'Duck' Mehta was trained.[277] It flew 1630 hours and twenty five minutes in its life span and was disbanded in December 1946. The Flight initially had Auster 3 aircraft which gave quite a trouble, in May 1945 the flight lost two aircraft in accident with no fatal casualties, the refresher courses it conducted had officers strength varying between two-four pilots 'No.2 course both of them {Captain Morris and Lieutenant Bromwick} arrived at 656 Squadron and up to now {1st May 1945} have put up 20 hours carried out six live shoots

275 Ibid.
276 See Artillery pamphlet No.7 Deployment of an Air OP Squadron dated 22nd July 1944, AIR 339.
277 Air 23/2552 letter no 5008/1/air 1[a] 24th November 1944

which were not up to standard compared with previous course'.[278] Flight had no hangar initially and one was authorised in April 1945 thus storms created mad rushes to the landing ground, aircraft were double picketed. For photographic training the pilots attended course at photographic section at Poona. Within the refresher flight a pilot was given 60 hours of flying in six weeks. In October 1945 the Auster 3 were and were replaced by Mark 5 'three new Auster V were collected from Poona...we have at last got rid of two very old Auster 3 due for major inspections'.[279] The flight did not had the water and electricity facility in the hangar, they constructed their own cook house at flight line and were happy to mess independently, commandant artillery school made a room available to the flight to be used as photographic dark room. Flight initially had its office in the artillery school and later they shifted to the Royal Artillery depot, flight had the accommodation problem in which drivers slept at verandah later the flight constructed its own accommodation. Transport was another issue so was construction of all weather landing strip 'panic in the area when they were informed that a General was arriving in Harvard may have some results'.[280] Later A Flight of 659 AOP Squadron took over the role of training from it.

On 27 June 1947, Lord Mountbatten now Viceroy of India circulated a memo highlighting the division of military assets, according to it in the first stage all the Muslims and non Muslims units were to be repatriated from Pakistan and India by September 1947, Artillery units were to be repatriated by November. An Armed Forces Reinstitution Committee was formed to settle the issue of military hardware. The task of bifurcating RAF and aviation related equipment was assigned to the air force sub-committee headed by Air Vice Marshal Perry Keene Who later became RPAF first chief. There was only one Air OP Squadron i.e.

278 Ibid , monthly report of 1587 Refresher Flight dated 1st May 1945
279 Ibid,monthly report dated 1st October 1945.
280 Ibid,monthly report dated 1st April 1945.

659 Air OP Squadron. It was decided to split this squadron. One flight consisting of four Auster V aircraft was Pakistan's share, remaining squadron was to form part of Indian Military. **659 Squadron Royal Air Force Air Observation Post** was raised at Firbeck in Yorkshire, England on 30th April 1943. Its first officer commanding was Major E.O.V Prendergast RA and the squadron was equipped with Mark III Auster and a single Tiger Moth. Squadron supported 8 Corps against 16 SS Panzer Division,it also took part in operation Over Lord and Operation, Market Garden. Along with 658 Squadron; 659 Squadron moved to India in October 1945 in response to extensive rioting and unrest. The squadron was involved in AOP training, but more importantly supported frontier forces in Waziristan . In August 1947 on transfer of power, the squadron was stationed at Lahore previously it was at Peshawar. Its assets were divided between India and Pakistan on 14 August 1947.

Chapter Eight 1947-1 956

No. 659

Air Observation Post

Royal Pakistan Air Force

Indian civilization prospered and flourished around River Indus[281] before it was invaded by the Aryans who originated in the Central Asia . These Aryans or mountainous people thus gained a psychological superiority over the plain dwellers. With the advent of Islam in 6th century and later conquest of Persia and Central Asia from 712 AD onwards. It were Turks who established, fermented and expanded these militant tribes under one ruler and very first mountain empire emerged at Ghazna.[282] The seventeen or more attacks by Mahmood of Ghazna at the end of first millennium on India and more specifically on rich Hindu temples not only eternally demoralized the Hindus but also set the foundation of a Muslim sultanate at Delhi in 1100 AD, this sultanate later expanded and captured Bengal. The bloodline of this empire were the Afghans and Turks Muslims which formed the fighting backbone of the military and ruling junta. The invasion of Tamerlane in 1399 and before him the Mongols in 12th century both originating from the Central Asia once again engraved the racial superiority of the mountain dwellers. Tamerlane had specifically targeted the Hindu population thus the religious supremacy of Islam over the non believers remained intact. Mughul Empire was established by Babur of Ferghana in 15260 after crushing the Muslim Sultan of Delhi, {Sultan Lodhi was an Afghan}

---

281 Mortimer Wheeler *Early India and Pakistan to Ashoka*, {Frederick,New York,1959}.pp 94-95

282 ,Syed,Ikram. *Modern Muslim India And the Birth Of Pakistan*{Islamic Culture ,Lahore,2000}pp 61-66

Mughul empire later not only protected the native population from the raids of mountain tribes of Afghanistan but also adopted a more liberal approach towards the non believers with varying degree of freedom. Afghanistan did not existed as a country under Mughuls,[283] Kabul the main city acted as a base camp for the Babur. In 1738 Delhi was ravaged first by the Persian Nadir Shah and later in 1763 by the Afghan Ahmed Shah Abdalli who in true sense is the founder of modern day Afghanistan. He kept Lahore, Kashmir and Multan as provinces of his Afghan empire and from this point in history the cultural clash took a new shape in the form of religious animosity between the Afghan Muslims and the Punjabi Sikhs.

British came as traders on the eastern coast and founded Calcutta city, many other European nations most notably the French were also present in India. The name India itself creates confusion. The River Indus in Sanskrit is known as Sindhu, when the SindhuDesh came under the Persians in 500 B.C; the Persians in their phonetics instead of S use H thus Sindhu became Hindu and Greek further twisted it into Indica, Turk called it Hindustan, natives as Bharat and British into India. By 1800 A.D; British were the major military power of the India, they suffered catastrophic massacre in the hands of Afghans in 1839 and from this point onwards the focus was on the western frontier. There were over 560 princely states in India they were semi independent, some had their own army as well. These states contributed substantially towards the war expenditure financially but more importantly in terms of fighting manpower. There was a parallel chain of command for the army in India. One was purely Indian Army regiments having a fair proportion of the British officers the other being pure white British Battalions. The backbone of the British Indian Army were the Muslims, Sikhs and Rajputs.

283 Charles, Miller, *Khyber,The Story Of The North West Frontier*, {Mcdonald& Jane's London,1977} ,pp xiii

On independence, Pakistan share of artillery regiments were eight gun regiments, one survey battery and one Air OP Flight.[284] These eight regiments were grouped under three artillery groups, the survey battery and Air OP remained under command. Director Artillery. The artillery directorate had five sections, in which the section No. 5 used to deal exclusively with anti aircraft and Air OP, Brigadier Harris was the first director followed by Brigadier Morley in January 1948.

The 1 AGRPA (Old 7 Division Artillery) Headquarters Rawalpindi had.

a. 1 Mountain Regiment Artillery.
b. 2 Field Regiment Artillery.
c. 3 Self Propelled Regiment Artillery.
d. 8 Medium Regiment Artillery.

At that time there was no Pakistani gunner pilot trained to take over the flight.[285] Those available were all British officers from Royal Artillery. Captain Azmat Baksh Awan a Muslim aviator was undergoing Air OP traning in England as part of course serial 61, he had opted for Pakistan. Major PD Morris thus became its first flight commander with Captain RD Raikes, Captain DS Murray and Captain GF Fox as section pilots. Initial technical and administrative manpower was as under :-

British Officers	- 4
British Army Non Commissioned Officers	- 6
British Army Other Ranks	- 28
RPAF Non Commissioned Officers attached later	- 7
RPAF Air Men attached later	- 11

The Indian contingent flew out to Amritsar on 7 September

284 Major General Fazal Muqeem Khan *The Story of Pakistan Army*, {Oxford,1963},pp,20-35.also see Major General Shoukat Riza *The Pakistan Army 1947-1949*{Services Book Club,Rawalpindi,1989}.pp-128-201. also see Brian Cloughley *A History of Pakistan Army Wars & Insurrections 2nd edition*,{Army Education Publishing House Rawalpindi,2002}.pp,2-9.

285 Azam,Cheema & Madni *History of Pakistan Army Aviation 1947-2007*{Army Press,Islamabad,2009}.pp-3-25.

1947. RPAF also had inherited ten Auster aircraft, these aircraft were lying at 101 Depot at Drigh Road Karachi. Later these aircraft formed the Communication Flight of RPAF.

The Aircraft.

Four Auster V aircraft inherited by Pakistan Army had the tail numbers, 228, 271, 377 and 625. It was a problem to restart a hot engine within the Flight as only three technicians could rotate the propeller and they were on great demand for out-station duties.

Operation Curzon: Waziristan Agency had remained a thorn in the rose of British Empire, the very first induction of aircraft in subcontinent is attributed to the unrest of tribes along the Durand Line since 1919. British had formed a Waziristan Area Command, with 30,000 troops force, which had its head quarters at Dera Ismail Khan, with brigade headquarters at Razmak, Bannu, Wanna and Gardai. The founding father of Pakistan Muhammad Ali Jinnah and the its very first Governor General popularly called Quaid-e-Azam intended to withdrew this force as a sign of reconciliation towards tribel code name Operation Curzon but there is another hypothesis that this force was pulled back to the new dimensions of security in the sub continent. The warring tribal lashkars attacked the retreating force and finally it were these tribes which invaded Kashmir that ignited the first war between Pakistan and India soon after independence. Warning orders for operation were issued on 01 October 1947. Squadron No. 5 R.P.A.F was already at Miran Shah.. On 11 October 1947 Flying Officer Siddiqui was given the first dual on Auster TJ337 and cleared for solo on 16th October in TJ 625. Siddiqui already had 300 hours on Spitfire, Hurricane, Harvard etc. he recalled {in October 2002}. 'I used to fly in support of army units, carrying out reconnaissance of road, delivering mail sorties. Miran Shah was the main control station, rebels on detection would be reported and soon Tempests of 5 Squadron would appear for strafing, army units would move from piquet to piquet, at times I would pick up casualty in Auster by makeshift arrangements and fly them to Peshawar.'[286] No. 659 AOP Flight was in the process of supporting boundary commission at Lahore when came an urgent requirement to provide support for Operation Curzon this seems bit crude as the biggest exodus in the history of mankind was taking place across the Punjab divided into two dominions by the Sir Cyril Radcliffe award bitterly criticized by the both dominions, the mass number of Muslims and lesser amount of Non Muslims walked across the line with marauding bands of humans slaughtering each other under the frenzy of religious zeal. Two sections were immediately dispatched to Razmak and a servicing station was also established at Peshawar. AOP Flight was to be used for inter communication, reconnaissance duties. Captain D. F. Murray and Captain D. S. Cox were the pilots at Razmak, whereas Major Morris and Captain R. D. Raikes stayed back at Lahore with flight headquarters and two sections (2 x aircraft) for supporting the Punjab Boundary Force. However flight headquarters and one section also moved to Peshawar on 7 October 1947.

In November 1947, the nomenclature of the flight was change from 659 Squadron into No.1 Air OP Flight Royal Pakistan Air Force the flight was entirely manned by Pakistani officers except for its flight commander, who was also relieved by Flying officer MM Jaffery in May 1948.

On 2nd November Captain Murray left the flight, F/O Siddiqui flew from Peshawar to Lahore via Chaklala to ferry back TJ228. Flights remnant at Lahore were supporting the boundary commission, Siddiqui also flew here 'our job was to fly in the Punjab and spot the refugees movement, then pass this information to Lahore for army cover, trains

286 Interview with Wing Commander Nazir Siddiqui, 2002. Also see Pakistan Aviation History, pg,3.

coming to Lahore had the doors sealed and lances put through the bodies, I would fly very low to assure the refugees, they were on the top of the roofs as well, road convoys coming from East Punjab were being harassed and attacked by the Sikhs, sight of Pakistani flag on Auster was always cheered by the refugees;'[287]

By 14 November Captain Cox also left and from RPAF came Pilot/Officer Steve Israel, Flight/Lieutenant Baig, Pilot Officer Mahmood. Technicians deserved special mention as they worked day and night to keep the aircraft flying worthy, they picked up very quickly from the British they included Younis, Amin, Khan, Azam, Afzal, Taj to name few. Siddiqui further recalls, 'There was no clear cut job defined, we all were motivated to work, work and work to make this country survive, I would fly in my own squadron as well as in Air OP. We would fly the air force crew from Peshawar to Razmak, fly political agents, and any other task given, Before flying from Peshawar we would inform the army units of our arrival and also get the weather, Auster had wireless to communicate'.[288]

Operation Curzon terminated on 31 December 1947, The flight flew 115 hours. On completion of the duty, the flight assembled at Peshawar, meanwhile Kashmir War had reached its peak , it would take 8-10 hours to drive from Rawalpindi to Kotli, 7 Division was commanded by Major-General Tottenham, he requested for Air OP during an informal dinner, he was then provided with it ,it took 30 minutes to reach Kotli in a fresh frame of minds.

Peshawar 1947-1948

In October 1947, RPAF station Peshawar was commanded by

287 Ibid
288 Ibid

Wing Commander Nazir Ullah, he had No 5, No 9 RPAF Squadron and No 659 AOP Flight under his command. The station meteorological section provided forecasting services for the operation staff at Chaklala and Risalpur as well. RPAF had inherited only a small Meteorological{Met} Section with main office at Peshawar, which was linked through a sluggish wireless telegraph {WT} with other airfields. Lahore had a civilian meteorological section

Air OP Flight remained at Peshawar till May 1948. The last section had also joined the flight from Lahore in January 1948. On 16 January 1948, Austers practiced formation flying, and on same day carried out formation flying in honour of Prime minister of Pakistan Mr Liaqat Ali Khan who visited the Peshawar air force base. On 19th they practiced cloud flying, on 9th February Auster VE601 force landed while on mail run. Air tests were usually carried out with the crew chiefs, Major Morris had special likings for instrument flying, strip landings and force landing practices. Air force officers like, Flying Officers Ramzan, Majid, Mallick, Razzak were there, apart from normal missions they would have separate training sorties for low flying and strip landings.

In March Flight Lieutenant Azam took over the flight temporarily, on 23 March Flying display was held in Lahore, the first in Pakistan by the RPAF. Austers represented the Air OP, they flew from Peshawar on 12th March and practiced formation flying ,joy rides were also given to the civilians .

Chaklala, Rawalpindi

On 24 March 1948 flight moved back to Peshawar and then moved to Chaklala on 4th June 1948. Reason of this move was multiple, first and foremost was the distance and time consumption in implementing the requirement of GHQ located at Rawalpindi, the least being the administrative inconvenience faced at Peshawar due to air force own expansion. At Chaklala flight was accommodated in a separate hanger with side offices. Chaklala was an army /air transport and para training school before independence, it had army units like Army Supply Corps. After independence It was short of manpower and gave a desolate appearance, metrological and fire services were almost non-existent and daily requirements were being met on an ad hoc basis. ***In November 1948, 659 Air OP Flight was re designated as No 1 Air OP Flight and stationed at Chaklala***.

Realising the future necessity, General Headquarters (GHQ) selected five gunner officers in October 1947 to undergo AOP training in UK. The officers were Captains Naseer Ullah Khan, Moin Ullah Karmani, Ghulam Jabbar, R.I.U. Khan and M.S. Wazir. But as luck would have it, out of this group, only Captain Kirmani and Captain Jabbar came back successfully they returned to Pakistan in September 1948.

Pilot Training at England was conducted in three phases.

Phase I	Pakistani students attended Artillery School Lark Hill, Gunnery procedures were refreshed, static observation post was used to practice Air OP procedures for shoot on ranges.
Phase II	Elementary flying training school R.A.F Boker Hill. Students were given 45 hours dual and 30 hours solo on Tiger Moth aircraft including 4 hours of night flying.
Phase III	Students were sent to No. 227 officers conversion unit Middle Wallop, {British Air Op School} here they

were converted on to Auster V and VI aircraft. In this phase flying was combined with artillery shoots, aerial photography, surveillance and other task required of an Air OP Officer.

Reminiscences

Brigadier Jabbar the senior most Pakistani AOP living aviator leads a retired life in his suburban house in Islamabad. He remembers those days, 'I was commissioned during World War 2 and served in Burma. On 14th August 1947, I was at Mardan serving in a Self Propelled Regiment when Pakistan became independent. I had already cleared my medical examination for the Air OP Course in May 1947 from Ambala and was later selected for the eleven months long flying course along with four other officers. We sailed for England which took 17 days before we docked at Liverpool.

England was damp, windy and rainy. Only Captain Kirmani had been in England before, in connection with an Ack Ack Artillery course thus he became our guide. We stayed at London for a week in a bed and breakfast hotel which was costing us around 17 Shillings a day. We were dependent upon the Indian Embassy for our pay and other administrative arrangements. Pakistan Embassy was yet to be established. Indian High Commissioner was Mr. Krishna Menon who later became Indian Defence Minister. There was an Indian restaurant by the name of 'Shafi', which we used to visit quite often. Our uniforms were still of the Indian pattern with Indian ranks and badges. It was a battle dress consisting of barathea cloth jacket and Van Heusen shirt with changeable collar. The buttons had the insignia of Royal Indian Artillery. We at our own had put on the Pakistani flag on the side arm. Once we went to Indian High Commission for collecting our pay, Mr Mennon was also there and he remarked on seeing the Pakistan flag and Indian buttons 'Young men you still have few buttons to change'.[289]

Pakistani students first went to Lark Hill, the Royal School of Artillery, located in Salisbury Plains, south of London. In the course they were taught artillery procedures and tactics. Captain Nasrullah (uncle of late Lieutenant General Akhtar Abdur Rehman) was the course senior. In the course there were twelve British officers and a few Indian naval officers also. Students were given independent single room and a British soldier as batman. Classrooms were near by but training area was bit far off and students would use bicycles for going there. 'We had one week break after the course which we used for travelling around. There was a town by the name of Wharton close to Lark Hill where we would go frequently for food. England was facing rationing in those days and only one egg was allowed in the breakfast, the restaurant was run by an old kind hearted lady, one day I requested for an additional egg and she replied "rules are rules, you have to go out of the restaurant and enter again, only then I can give you an additional egg".[290]

Their next stay was at Bokker Hill, a small town in Buckinghamshire County. Bokker Hill was mainly created by the RAF to take load off from Cranwell. It was RAF war time training institute. There was no runway and elementary flying and glider flying was conducted on grassy strips, course duration here was of four months. There were classes on principles of flight, navigation, airman-ship etc. Accommodation was in wooden huts; 'Our elementary flying training started on Tiger Moth, my instructor was Flight Lieutenant Depass who was a World War 2 veteran pilot. In Tiger Moth, communication was the major concern; instructor and students would talk to each other through Gospel Tube. There was no radio set in the aircraft even for

289 Interview with Brigder Jabar 2002-2003. Also see Pakistan Aviation History p-9.
290 Ibid

communication with ATC {Air Traffic Controller}. Pre flight briefing was very comprehensive and covered all aspects mainly to over come this deficiency of communication'.[291] All manoeuvres including barrel roll, stall turns, and loop were taught and practiced in pre solo stage. It was at this pre solo stage that other three Pakistani students were dropped from flying despite best efforts of instructors. Solo flying was done in Tiger Moth from the rear seat. Captain Jabbar and Captain Kirmani flew seventy hours in this phase. Final phase of four month flying was at Middle Wallop, which is located in Hampshire County close to Lark Hill. It was an army setup where commandant was a group captain from RAF. Main emphasis here was on the strip landings. There were two phases in this part of the course, in first phase, pilots converted on Auster Mark VI and in advance phase learnt artillery shoots and other tactical aspects of flying. Later they were awarded flying brevets in a graceful ceremony'.

Indian Air Observation Post.

Out of 550 Indian princely states Hyderabad was the biggest and the richest followed by Kashmir, both these states became controversial as in case of Hyderabad the overwhelming population was Hindu but ruled by a Muslim nawab and in case of Kashmir it reversed with ruler being a Dogra Hindu and population Muslim, at the transfer of power both states opted for the independence which was not welcomed by India and Pakistan and thus started an animosity in sub continent which has lasted to date. Captain E.K.H. Krishna an Indian artillery officer started his flying in Japan on Auster and later went to EFTS at Cliff Pypard in UK and OCU at Andover his account of flying in UK is similar to Captain Jabbar's account, Krishna highlights the lack of navigation aids and

291 Ibid

ground landmarks in UK 'give me a single track road running north to south, a broad gauge railway that cuts across well marked embankments villages with their own huge banyan trees'.[292] The Refresher Flight at Deolali after 1947 became Independent Air Op Flight with five Auster Mk 5 aircraft and two pilots to fly them,third pilot was soon posted in after completing his flying training in UK and two more pilots were sent there which were the last one as later all pilots were trained in the Flight something similar to Pakistan. For **Hyderabad Police Action** the flight moved to Sholapur with three Auster however the flight commander became a casualty of snake bite just before the D-Day. Krishna flew reconnaissance flights deep inside Hyderabad, he like other pilots carried extra fuel tank on the observer seat for extra range, the Auster were able to photograph the Hyderabad's fighters under trees {later they proved to be dummy} 'operation lasted less than a week, we flew to forward strips each evening prepared by trundling tanks over dry paddy fields, division commander himself a pilot whistled us up for continuous reconnaissance flights'[293] Indians were also using Set 62 for communication.

Kashmir

Kashmir, in Sanskrit, implies land desiccated from water: "ka" (the water) and shimeera {to desiccate}. Land locked state of Jammu and Kashmir {J&K} is situated in the northwest of South Asia subcontinent, India in the south and south east{Himachal Pradesh &Punjab}Pakistan in west, in east China{Sinkiang &Tibet}and Afghanistan {20 miles wide Wakhan strip in Afghanistan separates it from Tajikistan}. Total area is 222,236 square kilometres and population 4 million{80 % are Muslims}. Kashmir under a Sikh Maharaja opted to remain independent

292 Krishna wrote a paper of his memoirs named Auster Days of an Army pilot in India, which is preserved in MAF, however the magazine in which it was published is not mark
293 Ibid.

at Transfer of Power, where as both dominions Pakistan and India strategically, ideologically, culturally, economically and psychological wanted Kashmir, his Muslim population rose supported by Pathan tribes of Hindu Kush and anarchy spread. Maharaja acceded to India on 26th October 1947 which was conditionally accepted by Governor General Louis Mountbatten on 27th October, Indian Forces started landing at Srinagar and The First Kashmir War started.

Pakistan Army did not employed its Auster aircraft in the Kashmir campaign that lasted from November 1947-June 1948 however Indians did made use of their Austers, an Independent Operational Flight {IAF} was raised with five Auster Mark 5 aircraft and four pilots operating from Naushera where a strip was made 'routine shelling and counter bombardment was established on very gentlemanly terms... mealtimes were taboo'[294] Other strips used by the flight included Sambha, Jhunger, Akhnur and later Rajori which was a Dakota strip also, all these strips were within the mortar and machine gun range of Pakistan Army which accounted for at least one aircraft and one pilot 'previous pilots did some cheeky flying taking operational risks ending in pilot and aircraft casuality'.[295] Indian Austers were troubled by the Pakistani anti aircraft guns which on many occasions put a hole in them and forced them to change their flying tactics, flying was in mountains which also included crossing over at 9000 feet Banihal Pass, pilots also carried battle casualties in Auster by making use of a Pakistani Harvard's fuselage reportedly downed by the Indians. The Indian Air Op also faced the similar dilemma of the ground commander's discipline issue where the pilots attire was disliked 'corps commander a Sandhurst type disliked my flying get-up...my gunner beret, my fawn creepers and never discarded flying suit worn even on formal luncheon for him at

294 Ibid.
295 Ibid.

forward posts'.[296] Krishna also highlights the human nature of senior army officers especially their fear of flying, he wrote that his divisional commander would cancel a sortie with propeller's running for no reason, later the same commander died in a jeep accident, in another sortie the corps commander dropped his operational orders out of Auster and was lucky that his top secret orders bag strap just wrapped around his wrist, commander was trying to stop the freezing air flow through his thick bag. Krishna later commanded the Kashmir Flight and also the squadron, he talks very high of his pilots who 'all became senior brigadiers and major generals, one holds Maha Vir Chakar and five others Vir Chakra.

Captain Gerald Terence Leeworthy

Pakistan Army realised that sending officers to UK for AOP training was a costly affair and resulted in considerable drain on the national exchequer. The decision was made to bring in one instructor from UK to train future AOP officers within the country till the time own army instructors are not only trained but also have gained sufficient experience. For the purpose Captain Gerald Terence Leeworthy DFC (Distinguished Flying Cross) came to Pakistan on 12 September 1948 and besides his training assignment he also acted as an advisor on technical matters pertaining to AOP.

Captain 'Jerry' Leeworthy was posted as a Flying Instructor in British Air OP School, Middle Wallop. He happened to be the instructor of Captain A.B. Awan when he did his Air OP Course. Captain A.B. Awan was so impressed by his teaching techniques and sense of humour that on his arrival he recommended Captain Leeworthy's name to the

296 There is no such thing as a formal lunch in the field especially in war but the British Army culture adopted by the Dominions reflects the core difference between the RAF flying culture and army insistence on formality even amidst enemy fire.

authorities for posting in Pakistan. Subsequently on the request of Pakistan Government, he joined Number 1 Air OP Flight in September 1948. The issue of pay and allowances was the very first one to be raised rather Leeworthy had misplaced his luggage on the ship and that was his first worry, it was based upon this experience that regular flying pay was initiated for the army pilots in Pakistan. in 1950 it was authorised under a Pakistan Army Order (PAO)[297] army pilots started getting Rupees fifty per month which was later increased to Rupees one hundred per month.

Training in Pakistan 1948

Training pattern in Pakistan-was as under

a. Elementary flying training at RPAF Risalpur, it was of 50 hours on Tiger Moth.at that time RPAF was using Tiger Moth as the elementary flying training aircraft. Although RPAF had inherited few Auster aircraft on independence but due to constrain of instructors they were not used

b. Advance training of 100 hours at Chaklala, included conversion on Auster V, and various exercises. It was conducted in the No. 1 AOP Flight. R.P.A.F

c. Tactical exercises of 50 hours including 4 hours of night flying and conversion on Auster Mk VI again this phase was conducted in the AOP flight. Final check out was undertaken by the chief flying instructor{CFI} from RPAF

d. Selected officers were all from Royal Pakistan Artillery, induction of non gunner officers started well after 1957, Medical

297 Pakistan Army Order 1950,General Headquarters Archives,Pakistan.

was conducted at Chaklala where Combined Military Board was located in a small room.

First Pakistani Student

Lieutenant M Salimullah (later retired as brigadier) was the first army officer to have been completely trained in Pakistan he joined the United Services pre Cadet College, Almora on September 1, 1945 and passed through Belgam, Mhow before commissioning on 9th June 1946 from Officer's Training School, Bangalore. After completing training, I was commissioned in Maratha Regiment and despatched to Artillery School, Deolali. On completion of the Young Officers Course I was posted to 8 Indian Field Regiment Artillery. On partition I came to Pakistan from Deolali in the advance party of Artillery School Nowshera and was posted to 2 Filed Regiment Royal Pakistan Artillery. 'Soon we were pushed in Kashmir to counter the Indian threat in the Uri Sector.'[298] 102 Brigade had done well in blocking the Indian advance towards Muzaffarabad. Brigadier Moreley, Director Artillery visited and expressed his satisfaction on unit's performance. While departing he informed Salimullah of his selection for the Gunnery Staff Course (UK) but Salimullah was inspired by flying thus he opted for Air OP Course and landed in Royal PAF College, Risalpur. On completion of basic training he joined Number 1 Air OP Flight at Chaklala on 2nd December 1948 for further training. He recalls. 'I had a long briefing from Major A.B Awan, the flight commander, who briefed me on the training aspects. I also met Captain Leeworthy (who was going to be my instructor,) Captain Jerry Leeworthy was a brilliant flying instructor and I was fortunate to be his pupil. His knowledge of ground subjects was excellent. My flying training schedule was easy. Two

---

298 Interview with Brigadier Saleem ullah 2002-04, private papers of Brigadier Saleem ullah also see Pakistan Aviation History p-9. Also see Pakistan Army Journal 1995.

hours of flying, three hours of ground subjects and two hours of normal unit administration as a quartermaster'.[299] Captain Leeworthy was very particular about the air exercises and he always made sure that they are understood on the ground before going up in the air. The pre and post flight briefings were comprehensive. The final phase was tactical flying and this is where Leeworthy was at his best. He would pick-up difficult unprepared strips and would make Saleem Ullah land there. 'This gave me tremendous confidence. I had hardly completed my 200 hours of flying before participating in Exercise Cooperation, when Leeworthy selected a strip close to Rawalpindi on Peshawar Road. This was a disused road, slightly curved, hardly 100 yards in length, nearly four yards wide with tall trees on the western side. The approach was one sided coupled with crosswind. Leeworthy was a master in concealed approaches and take-offs and expected us to land on a 150 yards strip with high trees on both sides. During take-off, the wheels invariably scraped the tree top branches, but I learnt a lesson that only a good instructor would produce a good pilot. My flying with Leeworthy was of great help to me during my flying Instructor Course in UK'.[300]

Captain Leeworthy used to lay great emphasis on map reading. He maintained that a pilot must know the relationship of map and ground. Accordingly, he would conduct map reading exercise once a month. The flight was deployed with a spacing of approximately 1000 yards. They were given vary light pistols and cartridges with a timetable when they had to fire. The pilots were given a flight path 5000 yards away from the troops with different heights. When the pilot was positioned at the flight path, troops would fire and the pilot had to record the location. Initially the result were not very encouraging but soon we improved.

On 20 February 1949, Major AB Awan proceeded to UK to

299 Ibid
300 Ibid

undergo the flying instructor's course and Captain Leeworthy became the commander of the Flight.

Lieutenant Salimullah was awarded the much-coveted flying brevet by General Sir Douglas Gracey, then Commander-in-Chief of Pakistan Army, on **3 March 1949** at Chaklala. Captain Jabbar recalls the event "it was simple affair we had send the typed card to the chief through the ADC and he accepted it, definitely it was a matter of pride for the Artillery Directorate and they cooperated. Station Commander Chaklala was invited verbally along with the other officers, we had borrowed few chairs from the mess and army unit as well. After the ceremony tea and coffee was served along with the patties and pastries, which were made in the mess'.[301] It was a matter of great honour and prestige to have the Army Chief as the chief guest. It is probably the only example of and Army Chief coming to a parade which had only one graduating officer.

Captain PJ Blaker (later retired as lieutenant colonel and migrated to Australia) was the second Pakistani and first Christian officer of AOP also completed his flying course at RPAF College Risalpur and reported to the flight for further AOP training in March 1949.

Temporary Move to Dhamial.

In 1948, RPAF was in the process of relocating their flying squadrons. Under this plan No 6 Transport Squadron RPAF, located at Lahore was to move to RPAF Base Chaklala. AOP flight had to move temporarily to Dhamial till the time the army constructs its own hangar at Chaklala for which air headquarter allotted suitable space { present PIA domestic lounge }The AOP Flight vacated the PAF hanger to No 6 Squadron RPAF and move to an abandoned airfield at Dhamial. Which

301 Interview with Brigader Jabbar.

was constructed in 1941 during Burma campaign and was used as the para training center, it was abandoned in 1945. There were two issues, which had to be resolved. Dhamial was occupied by the Vehicle Sub Depot, which had to be vacated and funds had to be allotted for the speedy construction of hanger at Chaklala. Major AB Awan before proceeding to United kingdom for his instructor course, had prepared his work sheet methodically to resolve these. Major Awan had a forceful personality and knew the art of presentation. Not many could refuse his demands. Before he left for his course, both the issues were resolved. The vehicle Sub Depot was ordered to vacate Dhamial and move to Golra immediately. The Engineer in Chief was allocated the funds for the construction of a hanger at Chaklala on priority basis. In the absence of Major Awan, Captain G. Leeworthy was promoted as major and he took over the flight, an average of 90 flying hours were logged each month by the flight.

In April 1949, Flight temporarily occupied Dhamial, All unit equipment was stored in the mud barracks and vehicles were parked in open behind the aircraft pens {present 8 squadron}. Officers and men lived in Chaklala and used to travel to Dhamial daily for work. One platoon of Ministry of Defence Constabulary was deputed for security duties at Dhamial, mainly due to the fact that it was all open and theft cases occurred, the culprits were the nearby villagers. Luckily one Bellman hanger was available where the aircraft were parked In December 1949 the hanger at Chaklala was completed and the Flight moved back Flight also received four new Auster aircraft. From the old stock, two had been damaged In 1949 Lycoming engines were no longer available and the British modified the Auster MK V by fitting a Gipsy Major Engine. It was called Auster Mk V-D. RPAF Forward Repair Unit Chaklala carried out this modification in our case and on its completion Captain. Salimullah was detailed to carry out the air test he still has vivid memory of that air test. 'The engine was little heavier than the Lycoming engine. I took off for the air test, carried out the tests listed in the Performa and came back to land. The approach was all right but at the final check the tail would not come down even when the stick was fully back. I knew it was a centre of gravity problem. I made four attempts in succession but failed. On the fifth approach I had to stall the aircraft and land with difficulty. During post flight the technical officer found that the ballast weights, which had to be tied to the tail spring, were not attached. This made the aircraft nose heavy'.[302] The aircraft gave a poor performance and was this discarded.

On 1 January 1950, No 3 University Air Squadron of the Punjab was also formed which along with the No 1 AOP Flight and No 6 Squadron detachment remained the three regular flying elements of the Chaklala station till June 1954 when Flying Instructor School (FIS) was also shifted here (which later on shifted back to Risalpur in 1961) about this time No 3 Air Scouts Troop and No 7 MODC platoon were also raised. By January 1956, Central Gliding School (CGS). No. 10 MODC platoon, No 103 Air Logistic Depot, No 2 Wireless Observer Squadron were also shifted from Drigh Road Karachi to Chaklala.

University Air Squadrons {UAS} were mainly formed to impart and promote air mindedness among the college and university students thus creating a pool of reserved pilots for the air force. RPAF had a policy of having 20 out of 25 pilots from these squadrons in its academy. Instructors were from RPAF, students were issued uniforms and on completion of training they were given pay and other facilities. In 1950 once RPAF adopted Harvard as the basic trainer they gave their Tiger Moth aircraft to it. Air Scouts were for boys age between 14-18 years of age mainly emphasising on aero modelling where as gliding schools were for the art of gliding ,the main purpose was to promote and

302 Interview with Brigader Saleem Ullah.

attract youth in the flying carrier. Army pilots were also trained in these UAS. PAF Air Chief Marshall Anwar Shamim{1981-1987} also came from the UAS.

Flight Routine

The working routine in the Flight was very enjoyable. Flying in the early hours of the day followed by administrative work and games in the afternoon. In the evening there used to be table tennis matches. Major Awan was a good player and it used to be a pleasure to watch him play against his elder brother Izzat Awan who was a star all India player.

The flight had four sections and each section had to be out on an exercise for a period of four weeks. These exercises were very enjoyable. The section consisted of a pilot, one aircraft, four technicians, one jeep, and a three-ton vehicle with water trailer (normal composition). The aim of the exercise was to practice advance landing ground {ALG} in the operational area, prepare road reconnaissance report, select suitable bridges, get familiar with the operational photographs. These exercises gave tremendous confidence to pilots and they could fly low level to any required point without the map. On return to base a thorough debriefing was arranged for the flight commander. Captain Jabbar recalls 'One usual morning, after the first assembly parade when the flight marched off, Major Awan came to me and told me to prepare one Auster V aircraft to drop him at Multan. After flying for three and a half hours, we reached Multan which was a barren airfield in those days. Apart from Artillery Group Headquarters and an odd artillery regiment, very little military existed there. A private car was waiting to take Major Awan to the place he wanted to go. After refueling the aircraft I returned to Rawalpindi. A day later I learnt that the preceding night Major Awan had received a call from Multan informing him about the death of his father. The next morning he had flown to Multan to attend his funeral',[303] but he remained cool throughout the flight.

Ferry of Austers to East Pakistan.

The first big challenge, to navigate over a long distance, with just the magnetic compass and a map, came in the winter of 1949. The requirement was to ferry four Auster aircraft to the former East Pakistan. The aim was to set up an University Air Squadron at Dacca, It was winter of 1949,. Major Azmat Awan wrote 'I was then Officer Commanding 1 Air OP Flight Royal Pakistan Air Force (RPAF). The telephone rang. would you be at the Air Headquarters day after tomorrow? About noon. Yes, I will, I assured.

I arrived at the Air Headquarters on the day. I was ushered into the Air Chief's office. Air Commodore Mcdonald, RAF officiated. I remember I was awed by what seemed to be a long walk to the table where he sat at the opposite end of the hall.

"We have a mission for you. A pilot training programme in East Pakistan must be set up soonest. The Indians would not allow over flying of armed aircraft but have conceded safe passage of few light planes. That is where we need you".

"Four Auster aircraft are to be ferried to Dacca under your charge. We would give you ex-polish pilots who have fought in the Battle of Britain. Convert them and also tame them to fly light planes.{in the early days RPAF had almost 30 Polish pilots on contract to make up the deficiency of aircrew, most left afterwards but few stayed and rose to the rank of wing commander} A Dakota would accompany you to carry

303 Interview with Brigader Jabbar.

maintenance crew, enough fuel, some spares and the like". I have spoken to my counterpart Mukerji. He has agreed to your night stay at Agra. A second stop at Gaya may be necessary. Work it out".

"Agra is an Indian Air Force Transport base. There may be some night flying in progress, in case you arrive late. Also Agra onwards, you may expect strong winds. Almost matching forward speed of your aircraft. The station commander would have been instructed to supervise early morning Meteorlogical briefing himself so that you set off under favourable weather conditions". I asked. "Amongst the pilots, may I have one of my own, Captain Jabbar? I would feel more comfortable".

"Yest, do that … the pilots and Dakota will report in due course. And my staff is available for any further assistance. Good luck!" On 9th November 1949 Flight Lieutenant Muckowski and Trantsky reported. So did Flying Officer Lanky Ahmad, the Dakota pilot. Converted Muckowski on Auster V E604 in 50 minutes.

First Day 30th October 1949

Chaklala to Lahore	2.10 hours
Lahore to Ambala	2.05 hours
Ambala to Palam (Dheli)	1.30 hours
Palam to Agra	.50 hours

At the commencement of a flight, pilot would clamber up to the Air Traffic Control Tower, file his flight plan and receive Meteorological briefing then on, he is at his own.

For all four of them it was their first long navigational flying experience into India. Azmat was a Major with seven years service. On 10 November 1949, his flying hours totalled 479.45 and he had been in command of a flying unit for 17 months.

They flew in pairs. Jabbar in the first and Azmat would take off last to make sure that others were safely airborne. This came handy at Palam. They refuelled first two aircraft took off but Muckowski's aircraft was stubborn to start. Azmat went to help. In the process he lost time. Half way through the leg Palam Agra, darkness descended and soon it was pitch black. It was an eerie feelings to be cut off from the world where the whine of your engine was the only sweet music. Luckily, the shimmer of Ganges River down below was a boon and Azmat would not let go of it. Suddenly Agra beacon was sighted then some lights over the airfield, a modest complex. Night flying was in progress. Azmat tried to join the circuit but was pushed off he succeed at the next attempt, made a short approach and landed long on the runway. Unstrapped as soon as the aircraft was off the tarmac just in time before the next Dakota roared past him. The pilot caught him in his headlights and reported to the Air Traffic Control. Shortly a vehicle with the station commander in it arrived and escorted him to the dispersal.

They were well looked after by the Indian Air Force (IAF). 'I could see that dinner our Polish pilots were some what over-indulging in drinks and I feared that they would not wake up sober the next morning to fly. We had a long flight, were tired and wanted to sleep but on host's insistence were driven for brief visit to Taj Mahal....That remains a memorable interlude'.[304]

Second Day 31st October 1949

Agra to Amousi (Near Lucknow)	2.10 hours
Amousi to Babatpur (Near Banaras)	1.30 hours
Babatpur to Gaya	1.00 hours

Next morning the station commander was there early to check on the meteorological forecast, they had to wait several hours before the winds turned favourable and were allowed to take off. Lucknow was an industrial town and they were prohibited to fly over it. Pakistanis

304 Azmat Bakhs Private papers,

were told to detour and land at a small airfield just beyond at Amousi. However, this leg was somewhat long and at the close end of aircraft's endurance. Should the winds turned against; there was the risk that they would run out of fuel. So three-quarter way up a short strip was hastily prepared, marked with a "tee" for an emergency landing, and the map coordinated provided. Luckily, pilots did not need it.

At Gaya, they landed in failing light. 'We were met by a police officer and slept in a rest house'.[305]

Last day 1st November 1949

Gaya to Asonsol	1.30 hours
Asonsol to Dumdum (Calcutta)	1.00 hours
Dumdum to Dacca	1.45 hours

Asonsol was hazy with smoke due to the presence of coal fields. Dumdum was their last leg in India. Azmat in Auster No. E612 checked out at the air traffic control as usual then signed some papers at the immigration. 'We carried no passport or visa. The formalities were minimal. They knew our arrival and the nest destination. We refuelled and departed, the leg to Dacca was difficult from navigation point of view. It had rained and the whole countryside at the height we were flying looked like one sheet of water. There were few landmarks. I unnecessarily worried that we may stray into Burma. Dacca had the tiniest of landing strip, half submerged in water and surrounded by few thatched huts. I landed first and the others followed. With some relief we ended our historic journey. We spent 4 to 5 days in Dacca and met the General Office Commanding (GOC) 14 division–then Major General Muhammad Ayub Khan. We took a flight back to Karachi in the comfort of a 'Constellation'. There instructions awaited that Azmat should ferry

305 Interview Brigadier Ghulam Jabbar{retired} Islamabad 2001 also see Army Aviation annual magazine 1997.

back a newly arrived Auster VI (F662) to Chaklala'.[306]

In retrospect, Air Headquarters did some excellent staff work behind the scene, 'I believe, we succeeded in our mission. We were given the task, essential direction and allocated resources. There was no hassle. The start point was Trust. And trust begets trust'.[307]

In 1950 G.H.Q issued training instruction which had following tasks for A.O.P

Observation, Reconnaissance, aerial photography, message drop and pickup, para drop,radio relay,aerial cable laying and ccasualty evacuation. This was the very first step in which the AOP's role was defined and it was not much different from the WW2 yet the there was no mention of transportation of senior officers other than the reconnaissance and it was this bone of contention which had marred the reputation and efficiency of the army, some how the other Pakistan Army still commanded by the British officers was not willing to grant this time saving facility to its senior commanders something which the Americans had long done so.

General Gracey's Last Parade.

Captain Blaker had completed his training way back in December 1949 and year after, Captain A.M.S Ahmed and Captain MM Karim followed suit. On 12 January 1951, General Sir Douglas Gracey awarded flying brevets to these three officers in an impressive ceremony. General Muhammad Ayub Khan, the Commander-In-Chief designate, was also present on the occasion.

306 Log Book Lieutenant General Azmat Baksh Awan.

307 Ibid

Captain Karim who had joined the Military Academy Dehra Dun in August 1947 and migrated to Pakistan on 14 October 1947, was commissioned in 2 Field Regiment and was later posted to 3 SP Regiment in 1949. He remained as ADC to C-in-C General Gracey from 1 April 1949 to 19 October 1950. Captain Karim had already flown 200 hours in India at Bihar Flying Club, Patna. He had served in RIAF and Navy as commissioned officer the only instance of an officer serving in all three services. At the end of his tenure with General Gracey, as per custom in vogue, the chief asked for his choice posting, on which Captain Karim requested for Air OP Flight, thus he underwent flying course from 20 October - 20 December 1950. At the graduation parade which was Gracey's last as the C-in-C, he consented to be the chief guest, more as an affiliation to his ADC.The other two officers who received the flying brevet that day were Captain Blaker and Lieutenant Ahmed.

In early 1949 General Headquarters selected Captain Percy Joseph Blaker an Anglo-Indian officer of 8 Medium Regiment to undergo flying training, he was known for his lively demeanour; he gained an instant popularity with the rest of the community. He completed his basic flying course at RPAF College Risalpur and reported to the flight for further AOP training in March 1949. It took him another year and half before he got his brevet. He was all alone in Pakistan as rest of his family was in England. He later rose to the rank of lieutenant colonel and commanded 2 Squadron apart from being the Aviation School Commandant. A cool, composed and an extremely popular commanding officer. He later left the army and settled down in Australia where he took his last breath in the company of his wife Irene in February 2003.

Lieutenant A.M.S. Ahmed was from Bengal. He joined Air OP Flight in June 1950 from 1 AGRPA where he was an intelligence officer

{IO}. He completed 14 hours of training with University Air Squadron and rest of training within the flight. He later got married to a medical college student and had two kids. Sadly he died soon in an air crash. His wife later completed her studies in medicine and raised both children, she then settled down in East Pakistan.

Brigadier Karim recalling that day narrates. 'We all three stood at attention, the C-in-C walked in along with few army officers. Some air force officers and civilians were also invited. After the brevet ceremony; the flight consisting of fifty five all ranks gave a smart march past. A demonstration of short landing and take off was also shown. Lady Gracey also graced the occasion and while having light refreshment in the hangar remarked." This place looks cleaner than a hospital'.[308] It really showed how the gunners and airmen worked to keep a place like hangar so neat and tidy

Durand Line

Pakistan and Afghanistan both have historical cultural, ethnic and religious ties that spread over a thousand years. The very first Muslim empire established at Delhi in the 11th century was created by the Afghans and in different forms they remained in power or as king makers till the creation of Pakistan, these two countries despite having such a strong bond never had good relations.[309] One major irritant is the Durand Line the boundary between the two countries that was drawn in 1893 between the British India and the King of Afghanistan and it remained in vogue till the creation of Pakistan in 1947 after which Afghanistan raised the issue of Durand Line.[310] The Durand Line extends approximately 1200

308 Interview with Brigadier MM Karim, 2009, Islamabad.

309 Ali ,Mehrunnisa, Ed, *Readings in Pakistan Foreign Policy 1971-1998* {Oxford,Karachi,2001}pp,330-332.

310 The American Institute of Afghan Studies "The Durand Line: History, Consequences and Future."

miles[311] from the Sarikol range in the north to the Iranian border in the southwest. Durand Line has scattered tribes living in the area since long, scarcely populated yet heavily armed. The provinces of Baluchistan, North West Frontier Province (NWFP) and Federally Administered Tribal Areas (FATA) are on the eastern side of the Durand Line and form the western border of Pakistan. On Afghanistan's side of the Durand Line are the provinces of Nimruz, Helmand, Kandahar, Zabol, Paktika, Paktia, Nangarhar, Konar, Nuristan, and Badakshan.[312]

In May 1951, the 2/16 Punjab Regiment stationed at Chaman was commanded by Lieutenant Colonel Akhtar Malik (Later lieutenant general). The Afghans had never really accepted the Durand Line. One day the patrolling party of 2/16 Punjab was lured by Afghans into Spin Baldack, where they were invited for tea etc. As per tradition own soldiers left their arms outside, later on they were made hostage and then set free without their weapons. A tussle thus erupted in which exchange of fire also took place and situation became grim.Two aircraft of Air OP Flight were directed to move from Chaklala to Chaman in support of field formation. and carry out reconnaissance, aerial photography, direct artillery and tank fire.

On 12th May 1951,[313] Captain Karim took off from Chaklala, along with Captain Blaker and two crew chiefs. After 01:20 hours of flying, formation refuelled at Sargodha which at that time was a deserted airfield manned by a *chowkidar*. The pilots endorsed their flight plan in a register. Fuel was carried in jerry cans stowed in the rear of the aircraft. After flying further two hours refuelling was carried out at Multan which was a brick lined strip and fuel was brought from city by Burma Shell

---

Nov 2007. http://www.hollingscenter.org/Reports/07-2007_Durand_Line.pdf (accessed January 8, 2011).

311 MujtabaRazvi,*The Frontiers of Pakistan,a Study of frontier Problems in Pakistan's Foreign Policy*{National,Karachi,1971}p 143.

312 Muhammad QaiserJanjua. "In the Shadow of the Durand Line; Security, Stability, and the Future of Pakistan and Afghanistan". Naval Postgraduate School, Monterrey, California, p22

313 Ibid, Pakistan Aviation History, p-18.

on a telephone call. Auster-V had an endurance of two hours at eighty MPH and Auster-VI had 3.30 hours. Aviation fuel, 73 Octane was not available with Burma Shell and it used to provide required grading of fuel by mixing 100 Octane aviation fuel with 70 Octane in a suitable ratio. Shell always paid special attention to the army requirements. At Khanpur, Captain Karim's Auster developed engine problem which was cutting at high speed (it was later found that fuel filter was missing), due to this Captain Blaker took off early and Captain Karim went to Khanpur Railway Station to inform Chaklala Base on telegram regarding the problem. Later, a crew chief along with spares came on Auster to Khanpur and rectified the problem. Captain Karim then flew on Auster E617 for Quetta. On 16th May After necessary briefing at the divisional headquarters at Quetta the two aircraft were placed at Chaman. For next two days (17, 18) the senior officers carried out reconnaissance of the area. On 19th advance landing ground (ALG) was selected near Khojak Pass. Austers were fitted with steel metal plate under and back of the pilot's seat for protection against small arms fire. They {Captain Blaker and Captain Karim} normally flew at 2000 feet above ground level. Brigade Headquarters was located at Sheila Bagh.The pilots stayed at the RPAF officers mess Samungli.

On 1st & 2nd July 1951, Captain Karim with Captain Hamza of an infantry regiment flew inside Afghanistan and reported hostile activities at Shin Naisi Than, Agha Jan and Dobandi. They also observed a concentration of about 200 Afghan soldiers dug inside Pakistan territory. The information was promptly passed onto 9 Squadron of RPAF through the brigade headquarters. Flight Lieutenant Saad Ullah and Squadron Leader Zafar Choudhry took off in a "Fury" fighter from Samungli Airfield and fired on these entrenched Afghans. RPAF aircraft were in communication with Captain Karim on CN 348 set and were directed on to the target. This was followed by a ground attack in which the

Afghans suffered heavy casualties and withdrew. Both army pilots also directed own anti tank guns and artillery fire in the conflict. Brigadier Karim recalls, 'We had two types of radio sets with us, Set 62 was for communication with artillery and the set CN 348 was for communication with RPAF aircraft. We would pass the information on artillery net for artillery fire and on air force net for air strikes'.[314] During this operation Captains Karim and Blaker landed on unprepared strips on occasions to guide own advancing infantry about the enemy disposition.

Escalation-1951

Prime Minister Liaqat Ali Khan paid a visit to Quetta and Chaman. At Chaman along with brigade commander Brigadier Altaf Qadir and Wing Commander Zafar Chaudhry, officer commanding 9 Squadron (later C-in-C PAF). Prime Minister directed that Afghans should be taught a lesson. In the words of Captain Karim who was also present there with his Auster. 'Prime Minister made a punch and said "*in logon ko maza chakadu*" {Teach these people a lesson}. On 18 July 1951 Prime Minister held a cabinet meeting to raise a tribal *lashkar* of 20,000 for defence against Afghanistan. While Pakistan was busy at the western border, India amassed her forces at the eastern border. Although the build-up was going on since January 1951 but it reached the point where flight of two Austers was recalled hurriedly on 20th July 1951. On the way back Captain Karim's aircraft forced landed at Multan. 'I failed to locate Multan due to severe dust storm and forced landed at Shershah as I was running short of fuel'.

Flight was assigned the additional task of monitoring the blackout drills in Rawalpindi during this escalation. 'We would fly the commissioner who was a British, senior officers from GHQ would also be there, it was the

314 Ibid

fear and mental preparation for the looming war with India. The long six years of WW2 had left its marks on every one.' Recalls Brigadier Jabbar.

Premier's ride in Auster

In 1951 Mr Liaqat Ali Khan was touring Punjab although he had RPAF to look after his air transportation aspect, however he was at Gujrat where air force could not operate, therefore Lieutenant Colonel Habibullah who later became lieutenant general requested AOP to ferry the prime minister. Major Awan the flight commander had the honor of flying him in the army aircraft.

On 27 November 1951 Captain Saleem Ullah went to Bahawalpur to cooperate with 6 Bahawalpur Field Regiment. He came back with new ideas and new fire discipline orders. The few terms he learned during his stay at Bahawalpur included 'Latha Phairo' meaning Fire. 'D ,one, ka Latha phairo' meaning one round gun fire. 'Chir Chir Pai Hovee' meaning scale two. All shoots were successful and he wondered whether the fire discipline that emanates from Lark Hill is more correct or the one from Bahawalpur.

Royal Pakistan Navy

In October 1951, Royal Pakistan Navy was carrying out a study on the naval battle procedures which included development of an efficient target engagement system to reduce the engagement time, like the Air OP. The navy asked Army to provide a pilot and an aircraft to participate in the exercise. The army agreed and Captain Saleem Ullah was detailed to proceed to Karachi and take part in the naval exercise.

On October 13, 1951 at 0700 hours the pilots took off from Chaklala in Auster Mark VI (E654) with one technician (AC Malik). After 3.00 hours of flying the pilot reached overhead Multan and found the airfield covered with thick dust waste. Luckily, the pilot was familiar with the airfield layout, he planned to land from the north side, which was clear from obstructions. After making few attempts, he found gap and landed safely on the corner of the fair weather strip.

After refueling at Multan, the pilot took off for Jacobabad at 1200 hours. He landed at Jacobabad at 1500 hours. It was an abandoned airport but one assistant airport manager and his servant were present. There was no fuel available. The pilot had to go to the city to check on fuel. On contact he was informed that the vehicle has been repaired and is now on its way. The pilot and the technician had to spend the night on the runway. On October 14, 1951, after refueling, the pilot took off for Karachi and landed at Drigh Road at 1015 at 1015 hours. After parking himself at the PAF Mess, the pilot reported at the Naval Headquarter where he was briefed about the exercise.

The background of this exercise was that some time back a senior naval officer had witnessed a firepower demonstration at Tilla Ranges (Jhelum), which included a shoot by an Air OP. The naval officer was extremely impressed by the speed at which the target was engaged. He wanted to know if the navy can make use of the Air OP.

On October 15, 1951, the pilot carried out reconnaissance of Churna Island, which was the target area. It was a strange experience for the pilot. In the afternoon the pilot was given practice in the naval radio transmission {RT} procedures for communication with the ships and the passage of the fire orders, a communication set was fixed in the aircraft for communication with the ships on sea.

Between October 17-20, 1951, the pilot flew various sorties for various ships. The shoots were generally successful but used to take sometime before completion. Pakistan Navy practices firing on Churna Island every year and their opening round is close to the target but it usually took 3-4 ranging rounds to reach the target. Captain Saleem Ullah made the following observations.

a. Pakistan Navy had long range guns but very limited observation capability to make use of this extra arrange. They need a ship based aircraft/helicopter with complete navigational aids.

b. The ships while operating at sea pitch and roll while firing and this affects the gun data and the spread of the rounds on the target, inspite of the fact that naval guns are Gyro Controlled.

c. Found difficulty in fixing the flight path and the observation point since nothing was visible except water and Churna Island.

d. It was not advisable for a single engine aircraft to fly on sea.

As a last part of this exercise various discussions were held and the pilot returned to the base on 22nd October 1951.

On 21 November 1951 flight held its force landing competition, signals exercise was carried out apart from routine ALG exercises in Fateh Jang, Basal, Pindi Gheb, Campbellpur area each section had one pilot, one aircraft one jeep, one 2CWT vehicle, one operator communication unit, two drivers, two aircraft technician, one for the airframe, one for the engine.There was no exclusive cook etc, at times pilots would cook it himself, overall it was a great sense of unity in the section. At times problematic technicians would be taken along, as it was always a lengthy procedure to dealt them legally as they were under air force command channel at Chacklala, but in the field individual was sorted out through charge sheet by the end of the day.

At the end of 1951, corps exercise Centaur, the first of its kind, was carried out in area Pindi Bhattian-Sheikhupura. Two sections of AOP performed their duties, from 20th December 1951 to mid March 1952, in support of it.

Auster Aiglet & Tiger Moth

In 1952, the Auster Aiglet aircraft were procured. It was a commercial aircraft inducted for operational role. Its performance was similar to Auster V. There were two seats in the front (side by side) and two at the back. Its power/weight ratio was not satisfactory but these remained in service till replaced by L- 19 aircraft.

In 1953, AOP got one Tiger Moth aircraft from the air force. It was a biplane having an open cockpit and a skid as the tail wheel. The instructor would operate from the back seat while the student occupied the front one. It had no radio and no intercommunication system. Pilots were issued with two rubber tubes to be connected to a pipe for intercommunication between the two occupants. It was much more aerobatic than the Auster and proved useful for training. Major Saleemullah was categorized on this aircraft and he trained few pilots. Later on, the training continued on Auster V and Tiger Moth was returned to RPAF.

Aerial Photography

No 1 A.O.P Flight had the photographic equipment on its inventory, and could take vertical and oblique pictures. Corporal Raza from R.P.A F was attached with the flight. Major Awan intended to have photography from the aircraft since 1948, flight papers[315] revealed that in November 1948 it was delayed due to the non availability of rolls. Fatehjang was selected as the target; aerial photos from Auster were taken, developed by Corporal Raza. Although they were distorted but encouraging. next an operational area was selected, it took some time for obvious reasons, they were interpreted by ground liaison officer Niaz Ahmed at air headquarters Peshawar. These were shown to concerned commander. In the words of Brigadier Saleem Ullah 'we made a hole in Auster-V and Auster-VI aircraft. Camera lens used to protrude out side for oblique photography. Mosaic for 1 Corps was made, they were very happy to get these pictures. Chaklala had its own photo laboratory and photo section. Also Corporal Raza an expert in developing films, he was a good cook and man of little patience for slow students'. Later aerial photography was disrupted due to non-availability of films.

In the same context, one day Captain Blaker was sent to photograph two bridges, one near Jhelum city and the other at Marala. After photographing the first one, he flew south instead of going north. By the time he realized this error it was rather late he forced landed at Bhagtanwala, a disused airfield. After landing he wanted to inform the concerned authorities. But the nearest place having a telephone was Sargodha about 30 miles away. There was no transport visible for quite sometime so he asked a cyclist to give him lift to Sargodha. This was accepted magnanimously but with a condition that Captain Blaker rides the two-wheeler with his leg power, which per force was accepted. When asked later as to what happened, the reply was, 'Chum, it just happened but the long cycle drive has taught me an unforgettable lesson.'

March 1952. Flight took part in exercise Mercury and Hazard

---

315 Form 540 was the monthly report & return issued by RPAF see November 1948 form.

by supporting No. 2 AGRPA, exercise was conducted in Pindi Bhattian–Chauharkana-Gujrat area, General Ayub also visited the exercise area ,exercise was commonly known as 'Haphazard'.

The First Fatal Crash

On 2 May 1953, Captain Vajid Ali proceeded to reconnoiter an advanced landing ground area and on return crashed 4 miles from Chaklala, overhead his home he did a loop, his wife was standing on the rooftop and saw him crashed ,it was later revealed that he was not putting on the straps. Captain. Karim who was in the RPAF mess remembers. 'it was evening and I was in the mess when someone told me of crash, I reached the crash site on bicycle, by this time other officers including the station commander was there,Vajid was taken to the hospitals where he died, he is buried at RPAF cemetery Chaklala'. This was the first fatal crash which spread much gloom. It was followed by one more similar accident on 11 June 1953 in which Captain AMS Ahmed lost his life while carrying out local flying, just about 3 miles East of Chaklala.

Instructor Training

RPAF was following the policy of getting their instructors trained from England at Central Flying School {CFS} Little Risington located near Oxford. These instructors in turn trained the students at Risalpur. Pakistan Army also adopted this pattern and was given one vacancy every year at CFS. Major Awan availed the vacancy in 1949, Captain Saleem Ullah in 1950 and Captain M. M. Karim was the last one to undergo training in 1953. Brigadier Saleem Ullah narrates, 'It was eight weeks long course. There were few allied officers as well in the course. There were studies but the emphasis was on flying skills. Each flying instructor had two students. Initially it looked amusing to have the flying instructor acting as the student but it was really tough because he would create situation which I had not anticipated. The aim was to teach future instructors to be ready at all time in the cockpit for any eventuality and never ever lower their guard in cockpit with a student'. Auster flying was interesting; especially the short take-offs and landings. Chief Flying Instructor would fly with every student. Overall it was a tough course with little time to loiter around. At the end, a graduation parade was held within the class room in which school commandant awarded instructor insignia. After termination of course the qualified instructor would be attached with the school for consolidation training. Major Awan Captains Saleemullah and Karim all remained attached with the faculty.

Ack Ack

Co-operation with anti aircraft {Ack Ack} artillery was another regular feature Brigadier Jabbar recalls, 'at Multan it would become very monotonous to fly sorties of two hours daily for a fortnight, we would remain there for 15 days with No. 2 AGRPA, we had an air force officer with us, for majority of gunners it was there first interaction with the air force officer, with their peculiar cloth ranks, so one of the soldier walked to Pilot Officer Ashraf the flight adjutant and inquired 'should I salute you'.[316]

1953

On 6th January reconnaissance for advance landing ground strip

316 Interview with Brigadier Jabbar

was carried out in the west ridge area ,which gives fair idea of Rawalpindi city of 1953. On 12 January Lieutenant Colonel Gul Hassan later army chief was given low level reconnaissance experience. Year also brought in two new faces amongst the fraternity. Captain Muhammad Zaffar Khan performed his solo on 4th August 1953. Captain Zaffar retired as a lieutenant colonel and is currently pursuing a successful business in London.Captain Naseer Ullah Khan Babar earned his first solo on 8th September 1953. His instructor was Captain Mahmood who narrates 'Once I sent Babar to Sialkot area to develop a strip, later when I reached there I found the strip to be very small. Any how under ego pressure I landed with great difficulty. On inquiry he replied and showed that strip had another 200 yards of length available but it had a 90 degrees turn'.[317] Captain Babar left a deep mark of his boldness and courage wherever he went. He earned his first Sitara-i-Jurat as an OH-13 pilot in Chhamb area during 1965 War when he rounded fifty five Indian soldiers as Prisoners Of War and making them march for miles under their own officers. Incidentally during the entire episode Major Babar was not even carrying his personal weapon. He later commanded the first Rotary Wing Squadron (3 Squadron). He is also remembered for his assertiveness and dedication with which he led the aviation contingent in East Pakistan during Cyclone Relief Operation of 1969-70. He earned his second Sitara-i-Jurat during 1971 War while leading the attack of 111 Infantry Brigade. He rose to the rank of Major General. He later entered politics and served as Governor of NWFP, he has also been a Federal Minister for Interior.

In 1953, Captain Mahmood (brigadier retired) became the first army flying instructor to graduate from RPAF Flying Instructor School {FIS} Chaklala, he stood first in the course. The RPAF instructors included Flight Lieutenants Rahat Latif, Inam and Wing Commander

317 Interview with Brigadier M Mahmood, 2003, Rawalpindi

Siddiqui. The other students were Flight Lieutenants Sadruddin and Aslam Butt on Howard and Captain Mahmood on Auster. The FIS was located on one side of the hangar and on other side was the Air OP Flight. Pattern of training was same as in vogue at Little Risington. Routine included early flying followed by ground schooling. Final check out was conducted on Auster for Air OP pilots by the Air Force Instructors.

Life of the Air OP Days

Between 1947-57, Army pilots were dinning in the RPAF Mess Chaklala. Student pilots would also be living in the same mess or in present day Joint Services Mess. Major Awan got married in December 1950. Mrs. Awan recalls that, I came from Karachi in Freighter aircraft of RPAF and was warmly received at the airfield. Soon Awans were given a three room hutted accommodation on 14 Connaught Road. General Ayub was once introduced to the ladies in a party and when he was told that I am newly wed and living in a house, he got surprised and exclaimed that times have changed. I informed him that I am living with my mother in law on which he with a grin asked 'Is she good to you' and I innocently replied, 'It is too soon to say'.[318] Captain Jabbar was also married and was given a two room hut. Captain Saleem although married since 1946 but he brought his family a bit later. Out of the first three ladies only Mrs. Awan is still alive. Captain Karim remained bachelor for a long time so was Captain Kirmani. Captain Blaker was a lively and jolly fellow especially on week end nights, He got married to his sweetheart Irene an Anglo Indian working in the Base Hospital. There were four dinner nights in a week and once a month there used to be a guest night in which married officers along with their lady wives were also invited. Customarily newly married officers along with their lady wives were

318 Interview with Mrs Lieutenant General (Late) A.B. Awan, Rawalpindi 2004

dined in or dined out in the mess. The new building came up in 1955. Incidentally the first Mess Secretary Major Karim was from Air OP and the very first party held in this mess was of Air OP. Major Karim's wife who later designed the Army Aviation Brevet was chiefly responsible for the interior decoration of the mess.

Games were compulsory, officers used to play hockey and football with the troops. Air OP Flight had the best hockey team in the whole garrison. Cricket had yet to gain present day popularity. In the indoor games squash and table tennis were very popular. There used to be friendly matches with the air force pilots which invariably Air OP used to win. However, they were good in squash. There was a cinema hall in Chaklala run by 501 Workshop. Officers would send their batmen to inform the manager of their arrival and seats would be reserved for a ticket of Rs two for the box. Dress code was quiet strict; it was unheard of to see an officer going to city without a tie. Felt hat was also very popular. Jeans had yet to be introduced in the world fashion scene, neither *shalwar kameez* were there as they are today rather *chouri dar pajamas* and *sherwani,* Liaqat Ali Khan style was popular. In winter *Sherwani* was a common attire. Mess waiters had yet to wear Jinnah Cap as part of their dress. Once a month there used to be a ladies night in the mess. Almost all ladies were good in cooking and other artistic hobbies, for instance Mrs. Awan had been a Company Commander of the Women Guards in 1949, her other contemporary was Miss Nusrat, who later became the first lady of Pakistan and even in that capacity always maintained her old comradeship with Mrs. Awan. There was no television or glossy magazine to pass time, even telephone was a rarity and all good or bad news used to travel via telegram, letters or through ladies.

In the mess elite of the town were invited on special occasions. Army and Air Force Chiefs would invariably be there. Air Vice Marshall

Perry Keane was a regular member. Tobacco was an in thing of that era, the famous brand was Craven A which used to come in tin packing. Almost all officers had a radio and Radio Ceylon was very popular.. It was in this era that old 78 RPM Long play records which had the capacity of 5 minutes recording were overshadowed by 33 1/3 RPM which had 20 minutes recording and playing time. Even in buses and private cars these record players were fitted. Alcohol was allowed in the mess and in private gathering but its consumption can be compared to today's smoking however there used to be a compulsory cutting in mess bill for alcohol breakage. Narcotics were unheard, except opium for medical reasons .Officers were getting 750 plus Rupees per month, most of it was spent in clearing the mess bills. Dress was another segment; the tailors at the 'Mall' did provide credit facility for suits etc. The club was another drain, however, all the bills had to be cleared by 15th of each month. Cheque bouncing back for any reason was a court martial offence. In 1954 many army officers were court martialled for this offence ,none from Air OP. Officers would buy the books and records collection was a favourite hobby. Cars were rare, there were very few cars at that time and people who had them were recognized as such with the car brand, however heavy motorbikes were quite a fashion. Air OP Flight had one black car for administrative purpose, its key used to be in the mess, although students were not allowed to use it except in emergency. Officers mainly used bicycles for transportation in which 'Raleigh' was the favourite. Married officers would hire a *Tonga* to fetch their weekly or monthly provisions from Raja Bazaar etc. The cinema halls on Mall road were frequently visited as they showed english movies. It was an era of Lawrence Oliver, Humphery Bogart, Orson Welles, James Cagney, Alec Guinness, Marlon Brando and James Dean to name a few. Indian movies were also screened in certain cinema halls. In 1951, it was a common sight to see General Gracey or lady Gracey driving their private car on private occasions.

1954

On 21 January, Lieutenant Colonel Tikka Khan later Army Chief was flown from Chaklala to Nowshehra and back. In February Exercise 'Gold Rush' of 7 Division was conducted, its reconnaissance sorties were flown on 7 February. In March Exercise Vulcan was conducted in Mianwali–Bhakkar area, Major Awan who was in Pakistan Military Academy as an instructor was attached with the flight for the duration of the exercise. On the afternoon of 15th March 1954, three aircraft took off from the ALG including Captain Nasir Ullah Babar, Captain Mahmood and Major Karim who was having Colonel Ashworth, the American Military Attaché with him on board. All three aircraft were caught in an unexpected dust storm. Captain Babar landed safely on a quickly chosen strip followed by Captain Mahmood who slightly damaged his aircraft in the process. Major Karim however landed at a curved road near Sarai Krishna. As he stepped out of the aircraft he was welcomed by all the senior officers including General Ayub who was standing there with the umpires of the exercise. On 16th March 1954 General Ayub, the C-in-C, flew in Auster-V{ E623} with Major Karim from Bhakkar Strip to Dera Ismail Khan where he visited Military Police Centre. He again flew on 17 March for reconnaissance of exercise area and landed at Mianwali Base whereas the aircraft returned to Bhakkar. In the words of Major Karim, 'It was important because by this General Ayub showed his confidence in Air OP Flight and praised the flight for good handling in bad weather'.

The year 1954 received two young subalterns from Artillery, Lieutenants Asad Ullah Khan and U.A. Isani, for flying training. Later on Lieutenant Isani developed Asthma and was boarded out in 1955. Lieutenant Asad Ullah later applied for newly raised Special Services Group (SSG). There are quite a few aviators who started flying after their stint in SSG, but Lieutenant Asad is the only aviator who left flying to don the commando jacket. He later went over to the civil service and retired as Commissioner Quetta.

The second batch of 1954 trainees included Captain Muhammad Khan (later retired as brigadier) and Lieutenants Faizuddin Brandon and Khawaja Shahbaz Ghalib (both retired as lieutenant colonels). These officers completed their course by end 1955 and were cleared for flying by Squadron Leader C.R. Nawaz, Chief Flying Instructor at the Flying Instructor School.

Aerial Reconnaissance for selecting first Armoured Division area was carried on 31st October1954, with Major General Afzal and Brigadier Faruqi on board, in general area Taxila, Attock, Fateh Jang, Kahutta, Mandra, on 2nd November 1954, reconnaissance was carried out in Dina, Kharian, Chakwal area.

In November 1954, flight participated in "**Exercise November Handicap**" which lasted till mid December in general area Jhelum, Kallar Kahar, Sohawa. It was a two sided exercise, the biggest of its nature in subcontinent till that time. Air OP Flight was spilt into two sections, Captain Zaffar and Captain Naseer Ullah Babar were attached with seven Division, west of Murid Airfield and other section was with 15 Division. During the exercise night flying was also conducted with improvised jeep lights along with artillery observation and aerial photography.

In 1955, Monsoon rains brought floods in Punjab, worst affected areas were Sialkot-Narowal sector. Sections of flight took part in flood relief operation. Aircraft also spotted the stranded people and passed messages for rescue.

Another batch of artillery officers comprising Lieutenants Mumtaz Baluch, Latif Awan and 2nd/Lieutenants Qayyum, Nazar and Hameed Abbasy arrived for flying training in mid 1955. These officers were trained by University Air Squadron on Tiger Moth (DH-82) aircraft after which they shifted to No.1 Air OP Flight for advance training on Auster aircraft. Number of new faces like Captain Riaz, Hashmi, Mustanir Trimzi, Saqlain, Rabbani, Nazir, Aslam Janjua, Raja Wali and Shafaat were on the list of new aspirants after the aptitude test. Aptitude test of officers was conducted by Major Jabbar, Major Karim and Major Saleem Ullah.

Artillery Ranges

The artillery shoot used to be an interesting event. Brigadier Saleem recalls the early days 'We had a regular allotment of artillery ammunition. The camp would be established near Nowshera Ranges. This used to be a healthy competition amongst the pilots as to who takes the minimum rounds for registration of the targets. The success usually fell to those who determined map reference correctly.Taking shoots from a suitable height was no problem but the engagement from low heights while flying behind a cover required a special technique'. The pilot had to select a flight path behind a ridge, fix an OT line{ a flight path which allows pilot to see the target and falling of artillery shell}, point on the flight path and adjust his flight pattern, seconds prior to the round landing on target the pilot had to pop-up, observe the fall of shot, pass corrections to the guns and duck down on the flight path. The time to observe and pass the corrections was first few seconds and only the experienced pilots won this competition. In January 1956, the participating pilots took a bet with the commanding officer of 18 Field Regiment that they would need no more than three ranging rounds every time to reach the target. In the bargain, four bottles of a heady-brew were won, pilots who contributed to this feat were Faiz, Asad and Ghalib.

The Days Of Leather Jackets And Fur Lined Shoes

2nd Lieutenant Hameedul Haq Razaul Haq Abbasy {Brigadier retired} was one of the last one to join Air OP Flight before it became a squadron. In 1956, he came for the aptitude test as a second lieutenant. His reminiscences encompass the view point of a junior officer. 'In 1956 University Air Squadron was disbanded and Light Aircraft Training Flight was formed. It had DH-82 Tiger Moth as the primary trainer and we had to earn our solo on it before we could enter the fraternity of Air OP family. DH-82 did not have any self-starter neither it had any brakes. Moreover, it would land only on grassy strips. We the students had its photographs in our rooms and used to wonder how on this God's earth we are going to get solo. It was sheer dedication of my instructor, Flight Lieutenant Nisar ul Haq that I was able to do my solo. Although I nearly finished his flying career in one landing. I came out scratchless from that crash whereas he had to remain hospitalised for weeks. In post solo stage we were introduced to aerobatics in which almost every student over stretched the aircraft.

We were issued with an Irwin Leather Jacket, fur lined shoes and cold weather trousers as part of flying kit for DH-82. In Auster we would fly in normal uniform and jacket. During our flying on Tiger Moth we would go to Rawal Nullah.It was a nullah then and we will carry out unauthorized beat ups over it. Auster Mark VI had a smooth engine but still we were scared of flying it mainly because of its trimmer's adjustment during landing. Aiglet as compared to any other aircraft was a Rolls Royce and only flight commander would fly it. Once during artillery shoots at Nowshera Ranges, Lieutenant Nazar took off and swung the aircraft which hit the other two aircraft parked there, with this all aircraft of Air OP located there were out of action and we came to Rawalpindi by road. Last two weeks of the course would be spent on Nowshehra Ranges. One interesting and useful exercise was message pick and drop. In this a rope of thirty feet would be dropped from the aircraft window and then we had to pick up the message bag and retrieve it to the cockpit. In the same time period the engines of Auster Mark-VI were modified at Drigh Road, Karachi. We would go by train and air test the aircraft there, before flying back. Our route would be Karachi-Nawabshah-Jacobabad-Kashmor-Khanpur-Multan-Sargodha and Chaklala.[319]

In 1955, the US Government finalise the details of the military equipment to be supplied to Pakistan. This aid was commonly known as 5 Division Aid as Americans had pledged to raise and equip these divisions, it was as follows:-[320]

L19 Bird Dog Aircraft	-	62
OH-13 Helicopters	-	18
L-23 Command Aircraft	-	01

Raising Of Squadrons

In a conference held at General Headquarters it was decided to raise the following units in following order of priority:-

319 Riaz Ul Haque Brigadier,late. Interview June 2002.
320 Army Archives, General Headquarters, Rawalpindi.

No. 1 Air OP Squadron with its maintenance support units. The intention was that it would be on ground before the arrival of the said aircraft and that the surplus trained pilots can be absorbed. it would provide a base for future expansion.	1956	21L19 Bird Dog Aircraft
Air OP School with its maintenance support unit. For the training of pilots to make up the shortage in the Air OP units.	1958	15 L-19 Bird Dog Aircraft
No. 2 Air OP Squadron with its maintenance support unit.	1958	21 L-19 Bird Dog Aircraft
No. 3 Air OP Squadron with its maintenance support unit Reserve.	1964	18H13S Helicopters 5 L-19 Bird Dog Aircraft

Artillery Directorate.

Artillery directorate GHQ exercised complete operational control over the Air OP units. US Aid aircraft were expected to arrive in early 1958. These aircraft had to be absorbed in units to be raised. Post of General Staff Officer Grade 11 (GSO-II) aviation was sanctioned in the Artillery Directorate. Major M. Saleem Ullah was posted in this vacancy. This appointment was a great help in resolving the raising problems. In early 1956, the Directorate prepared a table of organization and equipment (TO&E) for No 1 Air OP Squadron. The TO&E included the air force personnel for maintenance of aircraft. The proposal had to be agreed by the PAF before the Ministry of Defence could approve it. The proposal was sent to Air HQ in the normal routine. The Air Chief Air Marshal Asghar Khan rejected the proposal and his remarks were that PAF was getting the American Aid and it was raising more units for which more PAF technicians were required. In view of own requirement, the PAF cannot provide the manpower to the Army. He further stated that PAF couldn't reconcile to the army plan of large-scale expansion programmed. The file with these remarks was returned to General

Headquarters the directorate got worried about these remarks.

In the directorate the urgency of the situation was discussed and it was decided that the case should be discussed with the Air Headquarters. Major Ashraf from Military Operations Directorate and Major Saleem Ullah from Artillery Directorate were detailed to proceed to Air Headquarters Mauripur to get the approval of the case. Major Saleem had good contacts with Air Headquarters Staff. The team called on Air Vice Marshal Yousaf, In view of air chief's remarks, he showed his inability to help. He however arranged a meeting with the Air Chief in which chief, explained his difficulties and his inability to support the Army proposal. He was of the opinion that the army should not expand its aviation activities so rapidly. It should go slow to enable PAF to provide technical support.

Major Saleem Ullah explained the army view points,[321] if the units are not raised. He suggested a compromise solution of the issue i.e. the PAF should accept the army proposal and endorse the file, that the technical manpower will be provided as and when available. This solution was discussed in detail and finally approved by the Air Chief. Team's missions was achieved. The Air Chief's arguments were solid and according to him no PAF technical cover could be provided for the next 3-4 years. This was the good indication to the army to make its own personal available for maintaining the aircraft.

The First Squadron Is Born

No. 1 AOP Squadron came into existence on 20 June 1956,[322] at Chaklala, under the officiating command of Major M. M. Karim.

321 Saleem ullah Brigadier,late,Interview September 2002 also interview with Brigadier MM Karim 2011 January at Islamabad.

322 War Diary 1 Air OP Squadron, Mangla,visited November 2010.also see History of Pakistan Army Aviation, p-26.

The promised machines took some time to arrive but there was enough strength of trained pilots to get the squadron into some shape. Lieutenant Colonel G Jabbar took over the command in January 1957 on promotion, after completing his tenure as a company commander at the Pakistan Military Academy.

Last Hurrah For The Auster

The Squadron continued flying old machines till the arrival of new American equipment. In October 1956 a requirement came to ferry two Auster from Karachi to Chaklala, which proved the last hurrah for the old birds. Major NUK Babar and Captain H R Abbasy[323] were dispatched for the purpose. They successfully accomplished the mission on 01 November 1956, which took a little over eleven hours of flying time with refuelling at various places en route. This was the second longest cross-country flight after the one undertaken over India back in 1949.

323 Abbassi,interview and his log book entry.

Chapter Niner

AIR OP AIRCRAFT

The eventual delays in raising of AOP squadrons are attributed to the RAF and non availability of suitable aircraft. AOP from the onset was conceived around the ability of small light single seat unarmed aircraft to quickly take off and land close to the artillery battery and this was based upon the flying experience of Taylorcraft an American machine being flown by the RA gunners in the RAFC, it was put to trial in Salisbury in 1938 and later a military version D type was flown in France but both were rejected by the Captain Bazeley, how ever the Taylorcraft had following parameters.[324]

Total useful load{Pilot and passenger and radio set}	360 lbs
Take off rest to wheels leaving ground.	195 yards
Landing run over 30 feet obstacles	200 yards fully braked
Climb to 1000 feet	2 minutes and 30 sec.
Dive from 1000 feet to ground level	19 sec.

In post Dunkirk the requirement of aircraft also incorporated to have an observer and later to have the armour plating, both these had adverse impact on the take off run of the aircraft and its manoeuvrability.[325] Stinson an American aircraft recommended by Bazeley was finally selected because it was already selected by the American Army and hundred such aircraft {Stinson O. 74} were ordered on 12th February 1941 but it never materialised thus army instead of concentrating on improvement of aerial observation through the existing aircraft pleaded

324 Lieutenant Colonel Bazeley's papers.report of D Flight in France dated 13th May 1940.Document No 19,Box No 5.MAF.

325 AIR CS 17382.'D.M.C dated 4th January 1943

'any aircraft' thus AOP settled for Taylorcraft Aeroplanes fitted with 50 h.p Lycoming engine which were ordered by Air Marshall Barratt to be fitted with Cirrus Minor engines with 90 h.p but this modification was dependent upon the delivery of Stinson aircraft but Barret insisted on the modification 'any forecast of aircraft deliveries from the United States had been shown by experience to be most unreliable',[326] later these words were proven true as American Army itself absorbed 90 such aircraft out of 100 earmarked for UK into their own air corps. The very first Squadron 651 was equipped with Auster Mark 1 and rapid changes incorporating army specifications resulted in delays, on 16th March 1942 the squadron was deficient of six aircraft finally on 19th March 1942 Air Ministry ordered production of 100 Auster aircraft thus six months were wasted in paper battle because army was not willing to listen to the advice of an air force officer, yet these aircraft from army point of view were temporarily and Tiger Moth were issued to AOP. This itself is an interesting feature as Tiger Moth was an ideal choice as it had no delaying factor and all AOP pilots were trained on it yet it was never considered for AOP duties because of its reduced visibility as compared to Auster yet the same Tiger Moth was used by the 656 Squadron Commander in Burma because of its better performance at high altitude, there fore had Tiger Moth been utilised for AOP instead of Auster from the beginning then these delays would have been avoided. In December 1942 MGRA agreed that 'Auster 1, 2 and 3 are temporary'.[327]

On 18th December 1942 Air Marshall Barrett forwarded a signal received from North Africa regarding the limitation of high wing aircraft and recommended a low wing aircraft preferably K.28/38 for service trials. The Vice Chief of Air Staff replied on 12th February 1943 in which he regretted the trial of K. 28/38 basing upon poor performance

---

326 Air Marshall Barratt letter dated 16th February 1943,reference ACC/S.81/Air 1.AIR 39/86.

327 WO 32/10413 Appendix X,Notes set before Army Command meeting at GHQ Home Forces on 13thAugust 1942.AIR ACC /S.14/26/1/Air dated 30th December 1942.

'apart from the relatively poor performance of the aircraft it would be quite imposiible to produce it in any quantity for many months by which time Auster 2 aircraft will be in plentiful supply and it will be modified for the rear ward view...there is another side to this we have reached definite agreement with War Office on the Auster they do not consider it 100% suitable but realise this applies to all aircraft...it is most disconcerting to the General Staff and our selves that you should continue to press for a replacement since this can only have the effect of destroying the War Office's confidence in the advice which we give them...may I ask you to accept the Air Ministry's decision in this matter and to assist us as much as you can by supporting the settled policy'.[328]

By 1943 the concept of having parachute in the aircraft was incorporated but only for the pilot and not for the observer or passenger, incorporation of cameras in Auster is another such example. There is nothing wrong in these changes as they were bound to happen but since from the concept around which these aircraft were procured was simple but with time it kept on changing there fore the delays are attributed to army and AOP. In the end Auster 6 was put to trial for casualty evacuation as well.

The wireless was a constant dilemma for the AOP, army sets were not compatible with RAF sets thus Austers had the No 9 and later No 21 sets, these were bulky sets and either set could be placed in the aircraft or a passenger. The frequency of engine failures in Auster was relatively low other than in one spell of 1944 when new American engine were installed. Over all Auster was a worthy aircraft but purely for the sole purpose of artillery observation.

All the Austers of Air OP were constructed at Rearsby near Leicester by one company whose original name was Taylorcraft

328 VCAS letter to Army Co-operation Command,dated 12th February 1943.AIR 39/86.

Aeroplanes Limited[founded on 21st November1938] later it was changed to Auster Aircraft limited. Taylor's original company had been acquired by William Piper, for whom he designed the E-2 Cub. They parted company in 1935, and Taylor went on to form the Taylor-Young Aero plane Company which produced the Taylorcraft.

Taylorcraft Model C.

Completion date (first prototype) 24 April 1939; first flight (prototype-civil registration G-AFNW) 3 May 1939. An improved version of USA civil aircraft dating back to 1934 it remained in production till July 1939. A 55 HP Lycoming engine, two seats side by side having maximum all up weight of 1200 lbs and a service load of 140 lbs.

Taylorcraft Model D.

Model C with different engine mounting, non aerobatic, initiated in October 1939 and first delivery on 28th November 1939.A 90 h.p. Cirrus Minor engine with same seating arrangement along with a wireless set, maximum service load of 490 lbs. and all up weight of 1400 lbs.

Basic aeroplane remained virtually unchanged it was a high winged monoplane with a wing span of 36 feet and an overall length of 22 feet 10 inches. The wings and fuselage were covered in fabric. All Austers were started by hand, pilot had the facility of wheel brakes operated independently by pedals, a small aero foil located under the tail plane operated by a lever beneath the pilot's seat was an effective trimming device .Auster had inherent stability. Fuselage was of steel tube welded together in a jig to form one entity, framework had no bolted or riveted joints whatever, wings had two main spars of spruce planks with a fairly thick high lift wings non folding in nature, it was routine to transport the aircraft by removing the wings and carry the aircraft in a three ton lorry, engines used were American or British, American engines were air cooled flat with four horizontally opposed cylinders forward and downward, British engines were also air cooled with four cylinder inverted in line.[329] Wireless was carried alongside the pilot and the switch operating the set from send to receive was placed as a push button on top of the control column, it required a thumb pressure from pilot to send.

The fabric of Auster in the Burma Campaign was a matter of concern, in September 1944 the B Flight of 656 Squadron had to abandon their Auster after one tenure of duty in Monsoon, later the commanding officer estimated that the fabric can last only four months under such weather conditions. The Wooden propeller and the Perspex canopy also needed replacement after similar period.

Auster Mark 1

It was initially flown in North Africa and much improvement was made upon the experience especially in terms of visibility from the cockpit and to performance in regard to short take-offs and landings. Taylorcraft Model D with minor modifications to meet service requirements, order placed in March 1942, it was almost blind, pilot sitting boxed in under the wings with almost no view aft and only a very poor view sideways on a turn.[330]

Span 36 feet, length 22 feet 10 inches, wing area 185 square feet no flaps, weight empty 894 Ibs, weight loaded 1400 lbs. engine 90 h.p. cirrus minor 1. Maximum speed 118 m.p.h. cruising speed 85 m.p.h., stalling speed 36 m.p.h, climb 800 feet per minute. Take off run [no

329 Unarmed into battle,p-143 also see fire by order ,p-180.
330 Unarmed into battle,p-144

wind]135 yards. Numbered produced 100.

Auster Mark 2

On 24th September 1942, Air Ministry & War Office accepted the Auster 2, an improved version of Mark 1 with airframe strengthened, use of 0-290 engine was decided in May 1942 but engine was not available until September 1942. Maximum all up weight of 1500 Lbs with service load of 530 lbs. and Lycoming engine of 125 HP.

Auster Mark 3

After non availability of Lycoming engine the Gipsy Major engine was decided on 21st August 1942 and prototype flew on 9th October 1942 and deliveries started in January 1943. It had 130 HP Gipsy Major engine with a maximum all up weight of 1550 lb and service load of 470 lbs, further modifications increased the AUW to 1700lb.The improvement was principally in take-off performance, addition of hand operated flaps reduced the take off run from 135 yards to 100 yards with a stalling speed of 40mph, rate of climb was improved from 860 to 1000 feet per minute, cruising speed was 90 mph.

Auster Mark 4

A redesigned airframe to have three seats, two side by side and third in rear but wireless set could only be carried at the cost of rear seat. First flight on 3rd May 1943 and delivery started in January 1944. 130 HP Lycoming 0-290 engine with 1750 lbs. of AUW and 604 lbs. of service load. Improvement was primarily in visibility, cabin perspex was extended backwards. Auster 4 came with an American Lycoming engine which had better fuel consumption. These engines gave a great deal of trouble in the early stages, this was due to overheating and restarting a heated engine was extremely difficult. Engine had originally been designed not for an aircraft but for heavy duty water pumps on American farms. This led to an indigenous modification carried out by 656 Squadron in the cowling thus allowing free air movement around the piston blocks.[331]

Auster V

Design confirmed on 15th March 1944 an up gradation of Mark 4 to allow communication and ambulance version. It had the same engine and AUW but service load was increased to 636 lbs. It was a three seat (side by side seats and one at the base) tail wheel aircraft with a very reliable American Lycoming engine. It could cruise at 80 miles per hour, had no inbuilt intercommunication, pilot had to strap an army No.62 set for communication with the guns, except the compass, it had no navigation aid. The instrument panel was basic. Instead of battery start, the propeller had to be swung by hand. Only a few technicians could swing, it had minimum lights but one could do the night flying. To navigate, it only had a compass and the pilot had to be extra vigilant to read the map, at night, a tiny green light would illuminate the instrument panel and a pair of red and blue lights would become visible at the two wing tips. Permission to land was indicated by a green flash on the heliograph operated by the air traffic controller. The floor of the rear cabin was strengthened to permit installation of third seat if required and provided ample space for fitting of F-24 recce camera. A two stage amber screening facility was provided for simulated instrument and night flying training. It was a problem to

---

331 Maslen Jones Fire by Order,p-49.

restart a hot engine. The Auster Mk V introduced a full blind-flying panel driven by a mechanised vacuum-pump to cater for bad weather flying on urgent communication duties. The auxiliary trimming surface below the tail plane was replaced by a standard elevator trimmer.

Auster VD

By 1950, Lycoming engine was no longer in production and in Pakistan replacement of Auster V had to be procured. Air Headquarters technical authorities recommended a change of the Lycoming engine and replace it with Jipsy Major engine. The modification was done at Forward Repair Unit (FRU) Chaklala. It was named as Auster Mark V D. The aircraft nose was heavy. Its spares were not available and no more aircraft was modified.

Auster VI

New version design to take 130 HP Gipsy Major engine and prototype flew on 25th May 1945, it had two seat in tandem, service load was increased to 688 lbs. and AUW to 2210 lbs. It was a twin seat (front and rear passenger seat), little heavier than Auster V, single control and had a battery start facility. It had cruising speed of 85 mile per hour, had the tail wheel, British Jipsy Major engine was more powerful then Lycoming engine. Conversion on this aircraft was a problem. The instructor sitting in the front seat would demonstrate few circuits and landings to the student sitting in the rear passenger seat, then change the position. The student had no problem in take off but on landing most of the students would swing. The instructor sitting in rear was helpless. The aircraft had the tendency to swing and students were shy to fly it.

Auster Aiglet

In 1952, the Auster Aiglet aircraft was procured. It was a commercial aircraft inducted for operational role. Its performance was similar to Auster VI. There were two seats in the front (side by side) and two at the back. Its power/weight ratio was not satisfactory but these remained in service till replaced by L-19 aircraft It should be borne in mind that the training version of Aiglet was two seat dual control aerobatic aircraft, Aiglet had the basic frame similar to the earlier J-5, but it had horn balanced rudder and elevators. Fuselage width was increased by four inch.

Tiger Moth DH-82

De Havilland's Tiger Moth was originally designed as a two seat light training biplane in the late 20's. It had tandem open cockpits, and the two wings were staggered and swept back. RPAF inherited some of them on independence and later bought 24 in 1948-1949.Army pilots were trained on it as elementary flying under RPAF later these were given to the university air squadrons where AOP pilots used to get the basic flying and then converted on to Auster in the AOP flight. These Tiger Moths were phased out from RPAF in 1957.

Technical Data

Maximum weight	1,825lbs
Span	29feet
Gypsy major engine air cooled	130hp
Maximum speed	109 mph
Ceiling	13,600feet

Epilogue

Apparently raising of few unarmed light aircraft squadrons seems to be a trifle affair in the context of second world war and probably this is the major reason that no dedicated work has ever been taken on this account yet this affair reflects the mind set of senior officers and enlightens us on how the war was fought or more precisely why British military fortunes suffered catastrophic debacles in this conflict, had it not been the material support of America the fate of the war would have been much different. Because if two services cannot agreed upon this minor aspect then it is futile to expect them to do great deeds. The root was in the political nature of the British military system at that time where strong lobby was necessary to have any thing done and unfortunately army after Kitchener never had a man of his stature.

Brigadier Massy's {Lieutenant General retired} efforts to revive the aerial observation of artillery fire were based upon his limited flying experience, to forward his idea he cloak it under the technical terms of artillery. Almost all gunners including Captain Bazeley believed in it more as a sign of regimental spirit. It was challenged by the RAF on professional grounds and from this point onwards it became a matter of ego for the gunners and army. The very trials of 1939 and 1940 at home and France restricted the concept to the adjustment of fire for field artillery only and keeping in view the national war effort it was a mere wastage of material and shipping space yet army pressed forward for it. Army out of reservation and knowing fully well that RAF will not agree to anything related to aircraft under army's control thus it never made the requirement clear and played with the domain of an elevated observer who will perform all the ground observers tasks. Army should have been more specific in its demand and requirements, it required aircraft if for nothing than at least for casualty evacuation for which RAF would had no answer.

AOP was a failure in operations when strictly judged from its constitution where it was meant only for field artillery. limited numbers of aircraft did not even allow to carry out this task appropriately ,it flew mere nine artillery sorties in its first campaign in North Africa yet false signals from artillery created an impression of its utility. It was a partial success when it is seen in a more broad spectrum of instant photography and rapid transportation of senior officers but than it was never meant for this. Fundamentally if there had been no AOP, its absence would have created no imbalance on the operations. AOP's real value came when they deviated from the rigid artillery role and acted more in the pattern of RFC, by carrying out the photographic missions and acting as an aerial transport for the commanders, the last duty is highly resented by the AOP pilots yet it was their solid contributions to the war. Another yard stick to evaluate the AOP is to compare its contribution as compared with the American Air OP's which on '8th May 1945 numbered 1380'[332] these Air OPs not only carried out their artillery tasks but also conducted multiple sorties including dropping of explosives for their infantry. It is a surprise that Major General Parham's picture is adoring the Museum of Army Flying at Middle Wallop along with Lieutenant Colonel Charles Bazeley as father of AOP, Parham's only contribution is a lone article on Auto Giro in 1933, Bazeley's war record is unimpressive, Lieutenant General Massy's portrait is surprisingly missing the very person who kept the AOP alive. Unlike American Army the British Army has never conducted an in depth study and evaluation of AOP and neither any academic has ever written solely about them thus an air of aura has been created by the lone historical account and couple of memoirs. Reservations of Air Marshall Arthur Barrett proved almost prophetic and are still valid today that an unarmed, slow aircraft cannot operate effectively from forward areas unless there is complete air superiority.

Observation from air through aircraft for army in general and

332 Report of the General Board United States,p.2.

artillery in particular was not a new idea as it had been practised in WW1,yet when it came into limelight in 1938 it was not entirely focussed on mere getting the observation for the artillery but with a suffix that it has to be done by the artillery officers only then it is useful, RAF point of view was that where as the importance of artillery observation is accepted yet that it should be done by them {gunners} is debatable and it was around this that army and artillery became stubborn and proved it through trials conducted in 1939 that it is only this option viable thus during the period under review the efforts of artillery remained to prove herself right, espirit de corps at its best. Army, artillery and War Office concentrated purely on demonstrating that gunners can fly and slow speed aircraft can survive the fighters yet they did not went into the tactical aspects of such venture and it was this lack of in depth knowledge, overwhelming espirit de corps, cumbersome staff work and procedures that resulted in the fiasco of D Flight in France yet on return instead of analysing the procedures and short comings War Office and Home Forces again got entangled in the air support for the army and trying to emulate the German army procedures in which Luftwaffe seems to be under army control. Purely from anti invasion deterrent the AOP were required as they could operate freely from the almost immense fields of Britain which provided cover both from air and ground. It was around this advantage that RAF seniors especially Air Chief Marshall Arthur Barret who himself was commissioned in artillery in 1910 and flew as an observer in WW1 as Chief of Air Staff CAS pointed out the inherent weaknesses of the concept. CAS observation were strategic in nature as he was concerned about the shipping space and cost these squadrons will incur vis a vis their utility. Even among the army circles the limitations of the AOP in terms of their limited duration of flight of only 20 minutes, height of only 600 feet and the reality that these aircraft have to fly in the rear of forward troops should have been enough to compel the khakis to concede to blues that 'lets improve the system' instead of rigidly adhering to the gunners idea. From artillery aspect these AOP were good enough only for the field regiments as they had a range of 12000 yards and are deployed 2000-3000 yards behind the front, the medium regiments with their range of 20000 yards are deployed around 3000-4000 yards behind, since these AOP could observe shells only till 6000-7000 yards thus they could support only field yet they had to fly within the flight path of medium and heavy regiments, resultantly in Sicilia and in European invasion almost 30% of AOP casualties occurred due to own fire. The limitation of height was imposed because of strong German anti aircraft fire and the time limit was accepted because of fear of Luftwaffe fighters. Thus with these limitations the whole concept of AOP flown by artillery instead of RAF became a flawed experience. These aircraft further required fighter cover for the duration of their operation. On a more analytically level the ability of a pilot to carry out artillery shoots in these twenty minutes remained a question which remained unanswer till the end. For this kind of effectiveness, it is pertinent that pilot is physically deployed with the artillery battery which was the original idea, yet the tactical consideration made it near impossible for this. with 1st Army this limitation was exposed to the full extent.

In Africa and in Sicily AOP did not rose to the occasion, only 10 artillery shoots were there yet artillery staff painted a false picture of their efficiency to an extent that further squadrons were employed, one major reason of this patronisation was the advantage which army commanders and their staff got from having at their disposal aircraft for transportation. One is not blaming them for the ride as it was also in relation to the task and work affecting the out put of the war but then these squadrons were not raised for this purpose. Even in Burma the resultant out put of 656 Squadron does not tally with the efforts that were required from the field formations in making the strips. This squadron evacuated only nine patients from the 'Admin Box' where as the American similar aircraft and organisation evacuated over 2000. The difference is starking.

The American adopted the similar pattern in 1941 when they authorised one aircraft as integral to every artillery regiment but they went many steps ahead by allocating these aircraft as integral establishment to armoured and infantry divisions and also allocating liaison and communication squadrons and flights to corps and army commanders, on the other hand till mid 1941 the debate within the War Office and Air Ministry was revolving around whether communication squadrons should be raised or not. later army commanders themselves forbid the use of AOP aircraft for transportation and staff termed these as 'personal taxis'. It was this stigma that has haunted the AOP pilots most as is seen from personal diaries and memoirs of AOP pilots and their squadron and flight commanders serving in Burma and Europe. AOP was a lucrative place during the war for gunner officers as most have highlighted the extra ration and leave available in RAF under pilot stress where as it was an alien word for the army, there fore many pilots were able to stay in AOP mainly due to the fact that senior pilots belonged to the same artillery regiments and had served together.

Selection of the aircraft for AOP was not resolved till the end of the war almost repeatedly it was conveyed and accepted that Auster aircraft is not suitable for the task, the original Stinson aircraft which ultimately was inducted by the American Army later proved its worth in Tunisia and Burma. Rota was the ideal platform for the artillery observation because of its ability to take off from rough fields ,had artillery persisted with it in 1935 the technical limitations faced at that time would have been overcome in couple of years but it was the obsession of aircraft that probably made Rota unsuitable.

Americans in fact totally revolutionised the concept of AOP by enlarging the canvass and inducting the idea of liaison in it. The use of the landing craft to get AOP airborne on 10th July 1943 at Siciliy is worth mentioning. The Canadians also had the AOP squadrons they copied the British system and had four AOP squadrons raised in 1943 onwards, their advantage was in their close liaison ,the AOP being integral to the Canadian Army and thus being able to support the corps and division. Where British system suffered most was the complex staff work and dual control of these squadron by RAF and artillery. For almost a year a complex legal battle was fought by the two services on the issues ranging from pay and allowances to vehicle marking and who is to write the report and who is to provide the ration. RAF had a much more rich menu than the army and its working environments more relax as compared the starched and stiff khakis. It was time and again the RAF which came to the rescue of pilots once they were put on mat by the army generals. In one instance a Captain Nicolson's remarks about the AOP ability to operate in mountainous terrain to commander 51 Division led a lengthy discipline correspondence and he was finally saved by the RAF who took the pilot's side, as this is what their culture is all about.

With the passage of years the same very Air Observation Posts Squadrons evolved into the present day Army Air Corps in United Kingdom and Army Aviation in Pakistan, now the aircraft and helicopters are under direct command of army chief and field formation commanders yet the tussle between the air crew and field commander is not over. These separate aviation corps have assumed the status of army air power and as such are reluctant to act as a mere transporter or supporting arm, rather the time has come where all other arms should be subservient to the air corps especially in present day war against terrors. Aviators are a different breed 'I guess we are considered ourselves a different breed of cat, right in the beginning. We flew through the air and other people walked on the ground:it was as simple as that' – famous American aviator General Spaatz[333].

333 James Tate, *The Army and its Air Corps, Army Policy Towards Aviation ,1918-1941* {Honolulu,University Press Pacific,2003},p,1.

Bibliography

Unpublished Sources

Pilots Notes General June 1941, updated in April 1943.

Report of The General Board U.S Forces European Theatre,Organic Field Artillery Air Observation, Study No 66,Reference No AORG No 9325.

Eye In The Sky, The Evolution of Artillery Air Observation 1918-1997 by Lieutenant Colonel John Gibson, War Studies 540, Air Warfare.Royal Military College of Canada, April 1997.

Brown With Two Shades of Blue, The History of B Flight in Normandy.

Note Books Of Air OP Course Students 1941, at Fire Power Museum.

Diary of a Section Pilot deployed in Tunisia, at FirePower Museum.

Private Papers of Lieutenant General {late} AzmatBaksh Awan, Islamabad.

Private Papers of Brigadier Saleemullah{late}, Rawalpindi.

Flying Log Books of Lieutenant General Azmat Awan, Brigadier M.M. Karim, Brigadier SaleemUllah, Brigadier GhulamJabbar.

Unpublished first draft of Pakistan Army Aviation History 1947-1997, Headquarters Pakistan Army Aviation Command, WestRidge, Rawalpindi.

Ahmed Manan. Many histories of Muhammad bin Qasim narrating the conquest of Sind,unpublished thesis for doctor of Philosophy degree in University of Chicago, Illinois, 2008.

Published Secondary Source

Andrew,Lyell, *Memoirs Of An Air Observation Officer* {Picton, Wiltshire,1985}

Anthony Farrar-Hackley Lieutenant General, *The Army In The Air, The History of The Army Air Corps.* {Alan Sutton, Washington, 1994}

Albert Manuq, *Artillery Through the ages, a short illustrated history of Cannon emphasizing types used in America* {US Government Press, Washington, 1949.}

Azam, Cheema, Madni, *History of Pakistan Army Aviation 1947-2007* {Army, Islamabad, 2009}.

Barber Noel, *War of Running Dogs, How Malay Defeated The Communist Guerrillas 1948-1960* {London, Fontana 1972}

Boog, Horst, Ed, *Conduct Of The Air War in The Second World War, An International Comparison* {Berg, NewYork, 1992}

Bowyer, Chaz, *The Royal Air Force 1939-1945* {Pen & Sword, Wiltshire, 1984}

Bidwell, Shelford, *Gunners At War, a Study of Royal Artillery in The 20th Century* {Arms & Armour Press, London, 1970}

Bidwell, Shelford and Graham, Dominick, *FirePower The British Army Weapons & Theories of War 1904-1945* {Pen & Sword, Yorkshire, 1982}

Blackburn, George, *Guns Of Normandy a Soldier's Eye View France 1944* {Constable & Company, London, 1998}

Cole, Christopher, Ed, *Royal Air Force Communiqués 1918* {Tom Donovan, London, 1969}

Cooling, Franklin, Ed, *Case Studies in The Development of Close Air Support* {United States Air Force Office of Air Force History, 1990}

Collier, Basil, *A History Of Air Power* {Weidenfeld & Nicolson, London, 1974

Cannon, Hardy, *Box Seat Over Hell The True Story of American Liasion Pilots and The Light Planes in WW2* {Association of Pilots, 1985, Texas}

Cloughley, Brian *A History of Pakistan Army Wars & Insurrections 2nd edition* {Army Education Publishing House Rawalpindi, 2002}.

Cooper, Malcolm, *The Birth Of Independent Air Power British Air Poli-*

cy in The First World War {Allen & Unwin, London, 1986}
C. Bayly, and T. Harper, *Forgotten Wars, The End of Britaiin's Asian Empire* {London; Penguin, 2007}
Dan, Snow, and Peter, *The World's Greatest Twentieth Century Battlefields* {BBC Books, London, 2007}
D. L. Fromow. *Canada's Flying Gunners, a History of the AOP of the Royal Regiment of the Canadian Artillery* {AOP pilots association of Canada, 2002}.
Forty, George. *British Army Hand Book 1939-1945* {Sutton, Gloucestershire, 1998}
Farrar, Anthony Sir General, *The Army in The Air History of the Army Aviation Corps*{Washington; Alan Sutton, 1994.
Fairbairn Geoffrey, *Revoloutionary Guerrilla Warfare* {Suffolk, Penguin, 1974}
Fuller, Charles, *The Second World War 1939-1945 a Strategical & Tactical History* {Eyre & Spottiswoode, London, 1948}
Gooch, John, Ed, *Air Power Theory and Practice* {Frank Cass, London, 1995}
Geoffrey Parker, Ed, *The Cambridge History Of Warfare,* {NewYork: Cambridge University Press, 2005}
Gjert Lage Dyndal. *Trenchard & Slessor on the supremacy of Air Power over Sea Power, Papers of the Royal Norwegian Air Force Academy, Volume 17* {Tapir akademisk forlag,2007}
Guillio, Douhet *Command of Air.*
Hughes, Ed, *History Of The Royal Regiment Of Artillery Between The Wars 1919-1939*{Brassy's, London, 1992}
Highman, Robin, *Air Power A concise History* {Macdonald, London, 1972}
Hyde, Montgomery, *British Air Power Between The Wars 1918-1939* {Heineman, London, 1976}
H. Wilson and J. Hammerton, Ed, *The Great War The Standard History*

of The All-Europe Conflict Volume 1, A new arm in the element by an expert in aviation {London: Amalgamated Press Limited, 1914}

Jackson Robert. *The Malayan Emergency and Indonesian Confrontation, The Commonwealth's Wars 1948-1966* {Yorkshire, Pen and Sword, 2008}

J. Loughitin, *Short History of the Royal Regiment of Artillery* {Aldershot, Gale & Polden, 1928}

Jones, Maslen, *Fire By Order, Recollection of Service With 656 Air Observation Squadron Post Squadron in Burma*, {Red Wood Books, Wiltshire, 1997}

Kennett, Lee, *The First Air War 1914-1918* {Free Press, New York, 1991}

Martin G Lieutenant General. *The History of 15th Scottish Division 1939-1945* {Blackwood, Edinburgh, 1948}

Mason, Tony, *Air Power A Centennial Appraisal* {Brassey, London, 1994}

Mead, Peter, *Soldiers In The Air, The Development of Army Flying* {Ian Allen, London, 1967}

Mead, Peter, *The Eye in The Air History of Air Observation & Reconaissance for The Army 1785-1945*{Her Majesty's Stationary Office, London, 1983}

Morrow, John, *The Great War in The Air Military Aviation From 1909-1921*{Smithsonian Institution Press, Washington, 1993}

Muqeem Fazal Khan Major General, *The Story of Pakistan Army,* {Oxford, 1963}.

Nagl Jhon. *Learning To Eat Soup With a Knife Counterinsurgency Lessons from Malay and Vietnam* {London, Chicago University Press, 2002}

Nesbit, Roy *An Illustrated History of RAF* {London, Salamander, 2002}

Parham and Belfield, *Unarmed into Battle The Story Of Air Observation Post* {Warren & Sons, Winchester, 1956}

Pearson, Michael, *The Burmese Air Campaign 1941-1945* {Pen & Sword, Yorkshire, 2006}

Ray, Cyril, *Algiers To Austria A History of 78 Division in the Second WorldWar* {Eyre & Spottiswoode, London, 1952}.

Riza, Shoukat, *The Pakistan Army 1947-1949* {Services Book Club, Rawalpindi, 1989}.

Riste Olav, Ed, *Western Security the Formulative Years 1947-1953* {New York, Coloumbia University Press, 1985},

Robinson, Rowan, colonel, *Artillery today and tomorrow,* {William Clowes & Sons, London, 1930}

Robertson Scot, *Development of RAF Strategic Bombing Doctrine,* {London, Praeger Studies}

Shennan Margaret, *Our Man in Malay* {Glouchestershire, Media Group, 2007}

Slim, William, *Defeat Into Victory* {Cassell & Company, London, 1956}

Sondhaus, Lawrence, *Strategic Culture and Ways of War* {New York, Routledge, 2006}

Tate, James, *The Army and Its Air Corps Army Policy Towards Aviation 1919-1941*{University Press of The Pacific, Honolulu, 2003}

Terraine, John, *The Right of The Line The Royal Air Force in the European War 1939-1945* {Hodder & Stoughton, 1998}

Wilfred Blake, *Royal Flying Corps in War* {Cassell & Company, New-York, 1918}

Magazine/Journals

Ray Sturtivant, Aviation News, 24 February-8th March 1984,

Journal of Army Aviation 1973.

Royal Artillery Journal April 1918-August 1920.

Royal Artillery Journal 1920.

Royal Artillery Journal July 1925.

Royal Artillery Journal July 1934.

Royal Artillery Journal October 1939.

RUSI Journal August 1933.

RAF Quarterly June 1944.

Pakistan Army Aviation Journal. 1995, 1996 & 1997.

Public Record Office

AIR	25/699	
AIR	25/722	
AIR	26/402	
AIR	27/1398	
AIR	27/418	
AIR	27/1242	
AIR	27/1711	
AIR	35/111	FOP Trials 1940.
AIR	35/113	FOP Trials 1940
AIR	39/69	Army Co-operation Command AOP Flying Training Septeber 1941-February 1943
AIR	39/70	43 Officer Training Unit Febrary 1943-May 1945
AIR	23 /5104	India Air Co-Operation Artillery 1943
AIR	23/2552	Control of AOP Squadron 1943-1945
AIR	23/4349	AOP Operational Reports 1945
AIR	39/143	AOP Policy
AIR	2/5251	
AIR	27/1245	
AIR	27/1241	
AIR	27/1245	

AIR	27/417	
AIR	27/418	
AIR	27/1242	
AIR	27/1398	
W/O	171/338	
W/O	171/132	
W/O	171/3958	
WO	216/77	AOP Squadrons formations February–March 1943
WO	205/21	AOP Policy July 1943-December 1944
WO	233/7	Division of responsibility AOP April 1944
WO	32/10413	Army Organisation AOP 1942-1943
WO	106/1599	Air Component for Field Force in France 1939 June-July
WO	204/1761	AOP Scales of Allotment April 1943-February 1945
WO	163/47	Army Council July 1937-October 1939
WO	315/5/137	Polish AOP Squadrons
WO	315/5/149	663 Polish AOP Squadrons
WO	32/13902	AOP Badge
WO	204/7535	Operation Husky. 654 AOP Squadron action August 1943
WO	233/60	
WO	373/36	Burma 22 March-19 April 1945
WO	373/37	Burma 26 April-24 May 1945
WO	373/38	Burma 21 June 1945
WO	373/40	Burma 12 July-2 August 1945

WO	208/1293	Notes on Japanese Artillery
CAB	79/11	Minutes of meetings 141-200, April - 21 June 1941.
CAB	92/111	
CAB	65/2	8 November 1939.
WC	75{30}	
CAB	75/2	
CAB	66/3/11	1939, Air requirement of army
CAB	24/280	
CAB	66/17/3	2 November 1939, lessons of Polish War
DR	30/42	Auster Aircraft

Fire Power Museum of Royal Regiment Of Artillery

MD/ 1996 Walby, Captain. RJ RA. *A history of B Flight 652 Squadron RAF entitled 'Beachhead to Bremen'.*

MD/3999 Brooks, T/Major. Peter RHA Typescript memoir entitled '*Peter Brooks Memoirs'* detailing his service 652 AOP Squadron RAF.

MD/4024 Creswell, Lieutenant Colonel. Jhon Hector RA OBE {1918-2004}

Collected pilots note books, course note book, manuscript diary of 7 Section, B Flight AOP Squadron RAF, pilots flying log.

MR 192 Report of the General Board United States Forces, European Theatre *Report on study of organic field artillery air observation*{study number 66} C. 1946

MD/2242 Bury, Lieutenant Colonel. Ralph OBE RA{TD}

'Brown with two shades of Blue' the reminiscences of a Air Observation 'Gunners' service with the Royal Air Force and Fleet Air Arm 1940.

MD/2794 Gibson, Jhon D. transcript entitled *'the Eye in the Sky' The*

evoloution of Artillery Air Observation Parts 1 & 2.
MD/2878 Neilson, Lieutenant Colonel, Ian G. DFC. *Transcript relating to the role of the Air Observation Post in Normandy 1944.*
Royal Artillery *Commemoration Book.*
Royal Artillery *Note No.15,15th April 1944.*

Museum Of Army Flying{MAF}
Short Historical Notes No. 654 squadron.
A Short History of No. 652 air Observation squadron 1942-1953 {part 1,1942-1945}.
Lieutenant Colonel Bazely's Papers.

Acknowledgement

This book is combination of two works which were undertaken at two varying time spans, the second part No 1 Air Observation Post Flight 1947-1956 was the pioneer work undertook in 2001 onwards when Major General Muhammad Azam General Officer Commanding Pakistan Army Aviation who happened to be my commanding officer in 7 Army Aviation Squadron 1990-1992 at Karachi initiated the much delayed project of compiling the history of own corps.

To know about something that happened almost half a century ago is a tiring cumbersome and interesting exercise. One man or officer was not present every where so how to piece the days by gone.Luckily the four senior most aviators were all alive like Brigadiers Jabbar, Saleemullah, Karim and Mahmood. Unfortunately Lieutenant General Azmat Awan had passed few years ago but he had already written a number of articles about his early days of flying which became handy. Brigadier Jabbar threw some light on his days at England, the ship journey, adventures of Captain Kirmani and the day other three students were dropped. These conversations were very interesting and absorbing. The topic would vary from the length of the strip, the engine power of a specific model to the menu in the restaurant and it would lead to the name of the lady running the shop. I noticed that Brigadier SaleemUllah did not smoke in his presence, later when I went to Brigadier Saleem Ullah, he just said 'I still cannot smoke in the presence of JabbarBhai, out of respect', all these events are history in it self. These two officers who were friend for well above half a century would always visit together. Contribution of Mrs AzmatBaksh Awan are worth quoting.She threw the light on the social side.

Brigadier SaleemUllah in coming days highlighted the pioneer days, ranging from the night he took Colonel Awan's car to take Captain Blaker on a date and vehicle failed to start next morning. His memory is so sharp that he even remembers the car number, which I have forgotten. Major General NaseerUllah Babar 'Bob' as he is known needs no introduction. He was the last one from Auster generation. He joined Air OP in 1956 and remained with Air OP till 1971. His name is synonym with bravery touching limits of irrationality [Brigadier Tirmizi's verdict]. How else can one define his capturing of Sikh infantry company unarmed.He was awarded SJ for this, the very first SJ of 1965 War and later in 1971 he was again awarded SJ in the same area for leading an infantry brigade attack from the front. He after retirement was appointed the Governor of Frontier Province. Bob in 1989 as Interior Minister again showed his characteristic 'devil may care' attitude while handling the hostage affair of school children bus in Peshawar. He put himself as hostage in lieu of children and later single handily overcame the terrorist. Bob has so many faces, his elephant memory, his recipes of Bar B Q, his humbleness and down to earth attitude.

After my retirement I went to University of Glasgow, Scotland to pursue a post master degree in war studies, the head of department Mr Simon Ball was instrumental in putting me after the history of Air Observation Post he guided me on the fundamentals of research; in the division of the subject. The historical archives at Kew, Middle Wallop, Woolwich, National Museum of Scotland at Edinburgh are rich and most friendly; in terms of documents of primary sources. Above all I am remorseful to my family and nothing can ever repay the patience they have shown in my this pursuit of knowledge.

Wing Commander Siddiqui, Brigadier SaleemUllah and Major General NaseerUllah Babar have all left this world on their final flight.

Made in the USA
Monee, IL
04 November 2026

67384086R00134